I0037176

Emerging Materials for Next Frontier Energy and Environment Applications

Edited by

Alagarsamy Pandikumar

Mani Alagiri

[1] Senior Scientist, Materials Electrochemistry Division, CSIR-Central Electrochemical Research Institute, Karaikudi-630003, Tamil Nadu, India

[2] Associate Professor, Department of Physics and Nanotechnology, SRM Institute of Science and Technology, Kattankulatthur-603203, Chengalpattu, Tamil Nadu, India

Published by **Materials Research Forum LLC**
Millersville, PA 17551, USA

Published as part of the book series
Materials Research Foundations
Volume 170 (2024)
ISSN 2471-8890 (Print)
ISSN 2471-8904 (Online)

Print ISBN 978-1-64490-328-5
eBook ISBN 978-1-64490-329-2

Distributed worldwide by

Materials Research Forum LLC
105 Springdale Lane
Millersville, PA 17551
USA
https://www.mrforum.com

Manufactured in the United States of America
10 9 8 7 6 5 4 3 2 1

Table of Contents

Preface

GO/rGO doped metal chalcogenide (CoSe, NiSe, and MnSe) nanocomposites for symmetric and asymmetric supercapacitors for energy storage applications
V. Ramesh, D.V. Sridevi, G. Bharath, G. Premanand, P. Umadevi,
N. Naveenkumar, A. Abhishek 1

2D Materials (2DM) based photocatalyst for environmental remediation
R. Roshan Chandrapal, G. Bakiyaraj 21

Recent advances in the electrode materials for electrocatalytic hydrogen evolution reactions
Thirugnanam Bavani, Sakthivel Vinith Kumar, Jagannathan Madhavan...................... 41

Role of biomass derived functionalized carbon as potential electrode materials for supercapacitor applications
S. Pavithra, K. Pramoda, A. Sakunthala, Rangappa S. Keri 60

Prospects of metal organic framework (MOF) as excellent electrode material for supercapacitors
Ariharaputhiran Anitha, Amalraj John... 94

Transition metal oxides-MXene nanocomposite:
The next frontier in supercapacitors
Jayachandran Madhavan, Pavithra Karthikesan, Harshini Sharan, Alagiri Mani 117

Synergistic effects in MXene: Transition metal chalcogenides to unlock supercapacitor potential
Harshini Sharan, Pavithra Karthikesan, Jayachandran Madhavan, Alagiri Mani 145

Keyword Index
About the Editors

Preface

The swift development of the global economy has led to the imminent problem of quick exhaustion of fossil fuels and the escalation of ecological damage. Consequently, there is a significant need for efficient and environment friendly energy storage sources. In order to address this issue, clean and sustainable ways of energy generation and storage are required along with removing the existing pollutants on our planet. Therefore, employing photocatalyst is an ideal way to counteract water pollution, as it is affordable, locks in the state of the pollutant, and can be utilised on a large scale. Electrochemical energy storage and conversion are often regarded as the most viable, environmentally beneficial, and sustainable sources of energy among clean energy technologies. Electrochemical energy storage technologies, such as batteries, supercapacitors, hydrogen generation, and storage, have been utilised in diverse sectors, including transportation, portable devices, and grid power buffers. To tackle clean energy sources, the electrocatalytic hydrogen evolution reaction is a very effective method for converting sustainable energy into a clean energy carrier, by water splitting. To attain the rising demand for higher energy and power densities in electrochemical energy systems supercapacitors play a significant role.

Henceforth, this book focuses on photocatalysis in environment remediation, electrode materials for electrocatalytic hydrogen production and supercapacitors. The recent advancements, state-of-the-art materials, synthesizing techniques, device fabrication techniques, the principle and mechanism of the working, current trends and future aspects will be discussed briefly in the forthcoming chapters. This book aims to clarify the practical aspects of comprehending and implementing technology in various industries. It also offers comprehensive technical information regarding emerging materials developed by field specialists that might turn readily accessible in the future. In addition, this book also addresses the technical difficulties to provide the readers an understanding of the practical constraints and their corresponding variables. We expect that university students, both undergraduates and graduates, as well as scientists and engineers working in the sector, will utilise this as a reliable source for their research and to generate innovative concepts for new devices that will enhance the technology's role in mainstream energy storage. By reading this book, readers will be able to effortlessly get up-to-date knowledge about electrochemical technology, including its basics and applications. We would like to acknowledge and convey our sincere gratitude for the work of all individuals involved in the design and creation of this book.

Dr. Alagarsamy Pandikumar *(Editor)*
Senior Scientist
Materials Electrochemistry Division, CSIR-Central Electrochemical Research Institute
Karaikudi-630003, Tamil Nadu, India

Dr. Mani Alagiri *(Editor)*
Associate Professor
Department of Physics and Nanotechnology, SRM Institute of Science and Technology
Kattankulatthur-603203, Chengalpattu, Tamil Nadu, India

Emerging Materials for Next Frontier Energy and Environment Applications   Materials Research Forum LLC
Materials Research Foundations 170 (2024) 1-20   https://doi.org/10.21741/9781644903292-1

Chapter 1

GO/rGO doped metal chalcogenide (CoSe, NiSe, and MnSe) nanocomposites for symmetric and asymmetric supercapacitors for energy storage applications

V. Ramesh[1*], D.V. Sridevi[2], G. Bharath[1], G. Premanand[1], P. Umadevi[1], N. Naveenkumar[1]
and A. Abhishek[1]

[1]Department of Physics and Nanotechnology, SRM Institute of Science and Technology, Kattankulathur, Chengalpattu, Tamil Nadu 603203, India

[2]Centre for Advanced Materials Research, Department of Chemistry, Rajalakshmi Institute of Technology, Kuthambakkam, Thiruvalur, 600124, Tamil Nadu, India

* rameshv@srmist.edu.in

Abstract

The scientific community is continually focusing on creating suitable energy storage devices and renewable energy conversion methods due to the growing demand for energy and the quick depletion of fossil fuels. Supercapacitors (SCs) are appealing energy storage devices with high power-density (P_d), good reversibility, and great cycle stability. They are a promising technology. Carbon-based materials like graphene, carbon nanotubes (CNTs), carbon nanosheets, carbon aerogel, and activated nanoporous carbon are used in these capacitors to store energy appropriately. Moreover, the metal chalcogenides (MCs) are a class of semiconducting materials that exhibit excellent electrical conductivity and possess a high theoretical specific capacitance. Because of their electrical and size-tuneable qualities throughout the manufacturing process, they have drawn the interest of researchers trying to increase electrochemical performance. Due to their distinct physical and chemical features, metal chalcogenides (MCs), sulfides, and selenides are highly regarded as promising candidates.

Keywords

GO/rGO Nanocomposites, Metal Chalcogenides based Supercapacitors, Energy Storage

Contents

GO/rGO doped metal chalcogenide (CoSe, NiSe, and MnSe) nanocomposites for symmetric and asymmetric supercapacitors for energy storage applications 1

1.   **Introduction** 2

    1.1   Renewable or clean energy sources 3

    1.2   Benefits of supercapacitor for energy storage systems 3

2.   **Principle, construction and working of supercapacitors**...4

    2.1   Principle of a supercapacitors..4

    2.2   Working principle of supercapacitors..5

3.   **Electrode materials for supercapacitors**...7

    3.1   Carbon-based electrodes...7

    3.2   Activated carbon..8

    3.3   Graphene...8

    3.4   Graphene-based composites 9

    3.5   Reduced Graphene Oxide (rGO). 10

4.   **Review of transition metal chalcogenides for supercapacitor applications**...........11

    4.1   Graphene-CoSe/rGO//AC composites based supercapacitors...........................11

    4.2   Graphene-NiSe/rGO//AC composites based supercapacitors 13

    4.3   Graphene-MnSe/rGO//AC composites based supercapacitors.........................13

Summary and overviews. 15

References...15

1.   Introduction

The world energy usage has increased dramatically in the last two decades and will only continue to increase. Renewable energy usage is projected to reach over 247 exajoules by 2050, up from 42 exajoules in 2000. The International Energy Outlook 2016 (IEO 2016) reference scenario predicts a substantial increase in global energy use from 2012 to 2040. The global consumption of marketed energy increased from 549 quadrillion British thermal units (Btu) in 2012 to 629 quadrillion Btu in 2020 and further to 815 quadrillion Btu in 2040. This represents a 48% growth from 2012 to 2040 [1]. Outside of the Organization for Economic Cooperation and Development (OECD), the developing non-OECD countries account for most of the global rise in energy consumption, with robust economic development and growing populations driving this increase. The demand for energy in non-OECD countries increased significantly by ~71% between 2012 and 2040. Conversely, in the well-developed, energy-consuming, and gradually expanding OECD nations, overall energy consumption increased by ~18% between 2012 and 2040 [2]. The world energy resources are the most significant amount of energy that can be produced using all available resources on Earth. There are three major types of resources: fossil fuel, nuclear fuel, and renewable. The Asian countries such as India, China, South Korea and Japan are most significant consumption of energy ~35% of the world. The above mentioned countries predicts a global production ratio of ~54.2 years for oil in 2012 [3]. The fossil fuels have a limited supply, and their extensive use is associated with environmental degradation. There are three main environmental problems occurs in the world, such as acid-precipitation, ozone depletion, and climate changes on a global level. Notably, tremendous increase in greenhouse gases (CO_2, CH_4) in the atmosphere and the increase of fuel prices are the major driving forces behind efforts to use renewable energy sources [4,5].

1.1   Renewable or clean energy sources

Renewable energy is obtained from natural sources that renew faster than used. The sun and the wind are two sources that are never exhausted. Renewable energy sources are abundant and all around us. On the other hand, fossil fuels, such as coal, oil, and gas, are nonrenewable resources that develop over hundreds of millions of years. When fossil fuels are used to create energy, they emit dangerous green-house gases like carbon dioxide. Renewable energy generates far fewer emissions than fossil fuels. Transitioning from fossil fuels, which now account for most emissions, to renewable energy is critical for combating the climate catastrophe [6]. Renewables are not only ecologically friendly but also economically profitable. They are cheaper in most nations and provide three times as much employment as fossil fuels. This positive economic component should encourage and promote the shift to renewable energy. For various reasons, several nations continuously increase their aspirations for renewable energy goals. For example, the European Union modified its 2030 mandatory objective from 27%, established in 2014, to 32% in June 2018. The latest objective included an item indicating that in 2023, nations would assemble again to deliberate on a revision [7]. The Indian government has set a challenging goal of achieving 175 GW of renewable energy by 2022, with 60 GW from wind energy and 100 GW from solar energy [8]. As the nation advanced, the Government of India increased the goal to achieve 227 GW by 2027 [9].

Over the past two decades, energy storage technologies have garnered much interest. Large-scale and small-scale energy storage systems are essential for developing and maintaining a technologically advanced civilization. Because of these exceptional qualities, supercapacitors (SCs) have garnered special interest in this context. These advantages include a greater power-density compared to secondary batteries, a higher energy-density compared to ordinary electric double-layer capacitors (OEDC), a longer cycle life, and improved environmental friendliness [10]. Due to their distinct benefits, they have been extensively used in portable electronic devices. However, because of their high power-density SCs are also gaining popularity as an attractive energy-storage option in situations where various power requirements are present, such as energy-storage for sporadic sources like the sun and wind or hybrid cars [11].

1.2   Benefits of supercapacitor for energy storage systems

Energy storage is crucial for maintaining a consistent supply of renewable energy to power grids, particularly during times of little sunshine or wind. Energy storage offers a means to attain flexibility, improve grid dependability and power quality, and accommodate the expansion of renewable energy (RE). It's not possible for energy storage systems (ESS) to work in all situations [12]. Furthermore, the requirements of emerging nations have often been disregarded. Developing nations often need more robust power systems. These are typified by inadequate monitoring and control technology, inadequate, unreliable, and inflexible power capacity, undeveloped or non-existent grid infrastructure, and a lack of maintenance. Energy storage can improve dependability in this particular situation [13]. When wind and solar energy are used, it may assist to replace expensive and dirty power produced from fossil fuels, while also improving the reliability of the energy supply. Most energy storage solutions are unsuitable for developing countries' circumstances and application cases [14]. Supercapacitors (SCs), a new type of electrochemical energy storage devices (ECESDs), operate similarly to rechargeable batteries (RCBs) and charge storage mechanisms (CSMs), unlike-electrostatic capacitors (UESCs) [15]. These devices are perfect for future electrical energy storage due to their exceptional performance. In terms of performance, SCs are positioned between rechargeable batteries and ordinary electrostatic capacitors since they have higher energy and power densities [16]. Over the last several years, there has been an increasing variety of uses for SCs. They are used as additional components in energy storage systems, such as electrochemical batteries, or as independent devices. SCs are a promising alternative electrochemical energy-storage technology to commonly used rechargeable batteries, particularly lithium-ion batteries [17]. An outline of the history of SCs technology and its current and potential applications is intended to be provided by this succinct point of view.

Recently, there have been big steps forward in electrical energy-storage technologies (EEST) like supercapacitors (SCs) and ultracapacitors (UCs). The energy-power curve shows that SCs are in the middle because they have higher power and energy levels than normal capacitors [18]. Energies and powers of commercially available SCs are usually between 4-5 Wh/kg and 10–20 kW/kg. New information, on the other hand, suggests that these numbers can be easily beaten soon. SCs have lower energy yields than batteries, but they have many benefits that batteries don't, most notably an almost endless lifetime. These materials also can charge quickly (unlike battery materials), have a wider temperature range, better safety, higher stability, and operate without any upkeep [19]. Instead of batteries, SCs work more like pulse power devices, and their discharge time is usually between a few seconds and less than one second [20]. SCs are now in high demand in transportation (about 70% of the market in the US), especially public transit like buses and trams, where they can be used as a stand-alone energy source or in addition to other energy sources. SCs are great for starting engines and can store charge during regenerative energy brakes. Also used in green energy (for example, to change the pitch of the rotor blades on a wind turbine), trucks, cranes, electronics, and home products [21]. The first mention of SCs was in a patent by General Electric in 1957 [22]. Then, from 1960 to 1970, the SOHIO Corporation invented devices that were similar [23]. Maxwell worked on the Ultracapacitor Development Program for the Department of Energy at the start of the 1990s and made SCs that could store 20 kW/kg of energy [24, 25]. These days, Batscap (Blue Solutions) sells goods in Europe that work just as well.

2.   Principle, construction and working of supercapacitors

2.1   Principle of a supercapacitors

A supercapacitor (SC) is an electrochemical capacitor with higher energy density and better performance efficiency than a regular capacitor. This is why the word "super" was added. Ions can be reversibly absorbed and released at the electrode-electrolyte contact, which is how it stores and releases energy. Nanomaterial-based SCs make the electrode surface area bigger so that they work better and have more capacitance as shown in Fig.1a. The porous electrodes consist of two layers with distinct material characteristics. The thicknesses of the two layers may vary as shown in Fig.1b. SCs and regular electrolytic capacitors (RECs) have many construction similarities. The system comprises two metallic electrode plates immersed in an electrolyte solution and separated by a thin insulating layer. SCs have a very high surface area on their conduction plates, and the distance between them is less than that of regular capacitors. A SCs is an electrochemical device that generates electrical energy without relying on chemical processes. Typically, the metal plates of a SCs are covered with porous materials like carbon or activated charcoal. The electrolyte substance between the two plates provides insulation and safeguards the device from charge leakage and short circuits. A SCs can store a substantial quantity of charge due to its capacitance, which is directly proportional to the enormous surface area of its conduction plates [26-28].

Additionally, the capacitance value exhibits an inverse relationship with the distance between the plates. This distance is much less in SCs compared to ordinary capacitors.

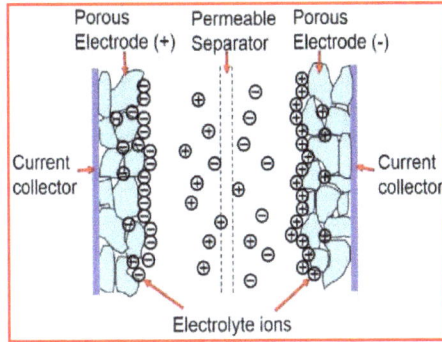

Fig.1a. illustrate of a SCs at the charged state

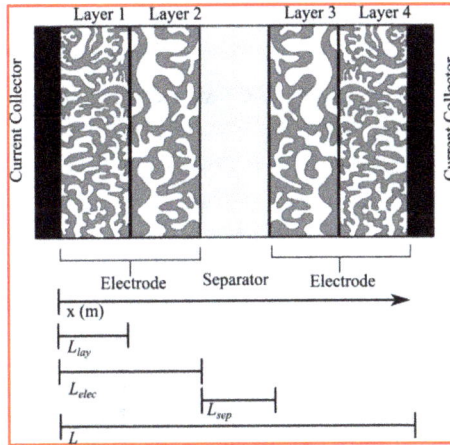

Fig.1b. shown the nano-porous electrodes consist of two layers with distinct material characteristics

2.2   Working principle of supercapacitors

The German scientist Heinrich Helmholtz (Helmholtz) first proposed the electric double-layer concept in 1879, which was the same idea that powers supercapacitors and electric double-layer

Emerging Materials for Next Frontier Energy and Environment Applications   Materials Research Forum LLC
Materials Research Foundations 170 (2024) 1-20   https://doi.org/10.21741/9781644903292-1

capacitors. Based on the fundamental principle of electrochemistry, the electrically charged double-layer capacitor works by creating a stable double-layer charge (electric-double-layer) with opposite signs at the interface between the conductor and electrolyte (liquid or solid). The electric double-layer charge cannot be drawn to an adjacent electrode that is opposite in sign, creating two electrodes for the real capacitor up to a certain voltage. With a double-layer electric charge that manifests as ions the size of a few nanometers, this de facto capacitor achieves an electrode distance on the nanometer scale since the electrolyte's electric charge is so small. The distance between the charge layers of capacitors (C_1 and C_2) are very small, resulting in a high capacitance value for each of them. The general relation of cell capacitance (C_{cell}) is [29].

$$\frac{1}{C_{cell}} = \frac{1}{C_1} + \frac{1}{C_2}$$
(1)

The EDLC (C_{dl}), at each electrode interface is denoted by

$$C_{dl} = \frac{\varepsilon A}{4\pi t}$$
(2)

where ε-is the dielectric constant of the EDLC region

is the surface-area of the electrode and

t- is the EDLC thickness.

The energy (E) and power (P_{max}) of SCs are measured according to following relations,

$$E = \frac{1}{2} CV^2$$
(3)

$$P_{max} = \frac{V^2}{4R}$$
(4)

The above is equation represents the relationship between the dc capacitance (C) in Farads, the nominal voltage (V), and R is the equivalent series resistance (ESR) in ohms. The capacitance of a device mostly relies on the properties of the electrode material, namely the surface area and the distribution of pore sizes [30]. The energy density is typically measured by the bulk capacitance of each electrode. This is due to the high porosity and low density of carbons. The voltage of the cells has a significant impact on both the specific energy and power of SCs. The operational voltage of SCs is contingent upon the stability of the liquid. Aqueous electrolytes, such as acids (e.g.,H_2SO_4) and alkalis (e.g.,KOH), provide favourable characteristics including high ionic conductivity (up to ~1 S/cm), affordability, and widespread applicability. However, these devices possess an inherent vulnerability that limits their voltage range, and their breakdown voltage is around 1.23 V. The specific-capacitance (measured in Farads per gram) of carbon materials with large surface area in aqueous fluids is often much greater than that of the same electrode in non-aqueous solutions. The reason for this is because water-based systems possess a greater dielectric constant [31,32]. In contrast, non-aqueous electrolytes have much greater electrical resistance compared to aqueous electrolytes. Consequently, capacitors fabricated from these materials often exhibit greater internal resistance. The power capacity and use of a capacitor are ultimately constrained by its elevated internal resistance. Multiple factors impact the internal resistance of supercapacitors. The inclusion of these elements leads to the determination of the value known as the equivalent series resistance (ESR) [33].

3.   Electrode materials for supercapacitors

3.1   Carbon-based electrodes

Most electronic industries are using carbon-based SCs. The electric-double layered capacitors (EDLCs) are made of carbon have many important features. Carbon electrodes resist corrosion and are inexpensive. They have good cycle stability, long life, and a working temperature range. They have strong electrical conductivity [34]. EDLCs power these supercapacitors. In charge storage, the double layer develops at the electrode-electrolyte interface. Capacitance depends on electrolyte ion availability in the surface area. Carbon-based electrodes accumulate more charge at the electrode-electrolyte contact due to their large specific surface area [35, 36]. Note that increasing carbon electrodes specific capacitance is difficult. Pore size, specific surface area, and surface functionalization must be balanced for the best outcomes. The active surface area and pore size distribution of the electrodes restrict the capacitance of carbon to a range of $0.15 - 0.4$ F/m^2 or 150 F/g. [37, 38]. Different nanostructures have pros and cons. Carbon nanotubes (CNTs) have remarkable qualities for rapid ion and electron transit due to their porous structure and good electrical properties. Unfortunately, its high production cost limits its applicability in many applications. Since recent advances, graphene, a two-dimensional carbon nanostructure, has shown promise in supercapacitor applications. High conductivity and specific surface area characterize graphene [39]. However, a high current density limits how much capacitance they can hold. This is because the conductivity isn't very high, and there are micropores [40]. To achieve, carbon nanostructures do at least two things: They help the composites have a high capacitance, and because they have high conductive, serve as a conductive path for electron transport. Different carbon nanostructures that can be used in EDLCs are shown in Table 1. These include carbon nanotubes (CNTs), AC, graphene, templated carbon, and carbon from carbides [41,42].

Table.1 Comparison of different nanostructures of carbon in terms of capacitance per volume.

Synthesized Material	Dimensions	Conductivity	Cost	Capacitance per volume
Carbon Nanotubes (CNTs)	1D	High	High	Small
Graphene oxide (GO)	2D	High	Moderate	Moderate
Activated Carbon (AC)	3D	Small	Small	High
Carbide derivative Carbon (CDC)	3D	Reasonable	Reasonable	High
Templated Carbon (TC)	3D	Small	High	Small

3.2   Activated carbon

Activated carbon is a prominent nanomaterial for commercial electrodes. They are chosen for electrodes due to their large surface area (3000 m²/g), moderate electronic conductivity, simplicity of manufacture, and inexpensive cost. Activated carbon can be created through physicochemical activation using wood, coal, and nutshells. Chemical activation involves activating chemicals like sodium and potassium hydroxide at 400 to 700°C temperatures. Physical activation refers to exposing carbon precursors to high temperatures (700-1200°C) in the presence of steam, carbon dioxide, or air. This exposure causes the oxidation of the carbon precursors. Activated carbon displays various physiochemical properties attributed to the carbon precursors and the activation processes employed. This results in a developed area of up to 3000 m²/g. Varying SCs activations generate varied topologies and pore diameters. Electrochemical features limit its operation, giving it a capacitance range of 150–355 F/g. Macropores (>50 nm), mesopores (2-50 nm), and micropores (<2 nm) are found in activated carbon [43]. Researchers have found a difference between AC-specific capacitance and surface area in various experiments. We got a low specific capacitance despite a big specific surface area.

3.3   Graphene

Graphene is a carbon allotrope. The material is composed of carbon atoms that are tightly arranged in a honeycomb lattice, with each carbon atom having sp^2 hybridization. Graphene is an ideal choice for the electrode material because of its remarkable features. Graphene may be created from graphite by oxidizing it to graphite oxide using either Hummer's technique or a modified version of Hummer's approach. Next, we employ a reducing agent, such as hydrazine solution, to decrease GO and produce graphene. The structures of eight different forms of carbon as seen in Fig. 2.

Materials Research Foundations 170 (2024) 1-20   https://doi.org/10.21741/9781644903292-1

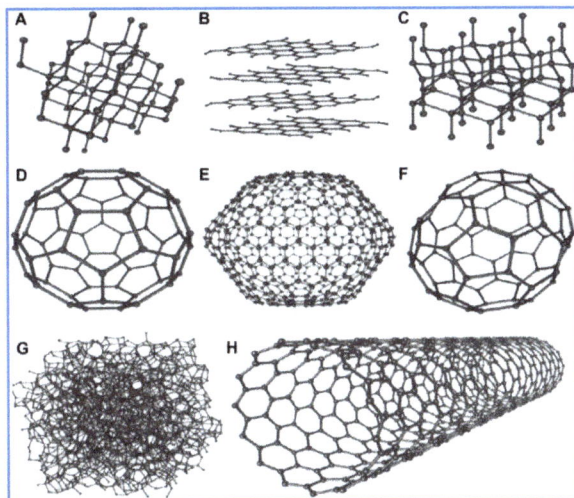

Fig. 2. shown the structures of eight different forms of carbon are as follows: (A) Diamond, which has a three-dimensional network covalent structure, and (B) Graphite, which consists of two-dimensional covalent plates. (Graphene is a single layer of graphite), (C) Lonsdaleite (D) C_{60} [0D, molecules] the compounds referred to as Buckminsterfullerene, (E) C_{540} Fullerene (F), and C_{70} Fullerene. (G) Carbon in an amorphous state, a single-walled carbon nanotube.

Graphene-based SCs have exceptional stability at elevated temperatures and in the presence of various chemicals, resulting in their prolonged lifespan [44]. Graphene possesses a significant specific surface area of 2630 m^2/g, a high thermal conductivity of 5000 Wm/K, a broad potential window, and excellent carrier mobility [45, 46]. These characteristics demonstrate that graphene has the potential to become one of the most durable choices for creating electrodes. The synthesis procedure of graphene oxide is illustrated in Fig.3.

3.4   Graphene-based composites

Composite materials refer to materials formed by combining many components with distinct qualities to generate a final product with exceptional properties. As stated before, graphene has several novel characteristics, any of which might result in exceptional composites. Graphene allows for composites with outstanding properties, improving the strength and conductivity of large-scale materials [47]. Graphene may be affixed to metals, plastics, and ceramics to produce composites resistant to heat and pressure. Graphene composites provide a multitude of uses, and much research is now being conducted to fabricate distinctive and captivating materials. The myriad applications of graphene composites in photo-catalytic processes are notable for their lightweight nature, versatility, and exceptional electrical conductivity [48]. Nevertheless, the authors will address the topic of composites made from graphene and transition metal oxides in the subsequent section.

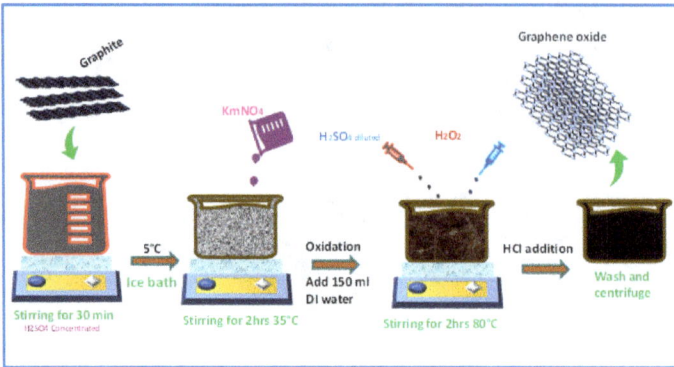

Fig.3 shows the synthesis procedure of graphene oxide (GO)

3.5   Reduced Graphene Oxide (rGO).

Graphene oxide is chemically or thermally reduced to rGO. The synthesis of rGO usually involves graphite sheets and oxidized graphene oxide is shown in Fig.4. The most common reducing agents for GO to rGO are hydrazine, hydrazine hydrate, l-ascorbic acid, and sodium borohydride. Plants and plant extracts are environmentally friendly and easy to manufacture, hence research has shown their use in rGO production. Plant extracts provide reducing and capping agents for nanoparticles. The functional groups of rGO oxides interact with DNA, proteins, peptides, and enzymes to aid chemical synthesis. Biomolecules are used to prepare rGO because they are cost-effective and nontoxic and biocompatible, requiring less labor and time for reduction. Therapeutic uses are possible because to low toxicity. These work because rGO has more surface area [49].

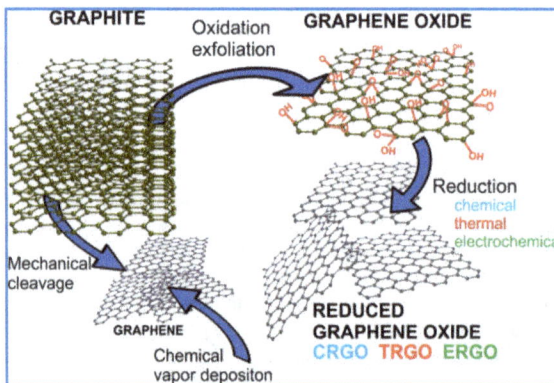

Fig.4) A reduced graphene oxide (rGO) synthesis process

4.   Review of transition metal chalcogenides for supercapacitor applications

4.1   Graphene-CoSe/rGO//AC composites based supercapacitors

Advances in energy storage technology have been made possible in large part by growing public awareness of the energy issue and environmental preservation. The benefits of SCs fast charge and discharge, extended service life, and large theoretical capacitance have drawn much interest [50]. The common co-precipitation method was employed the study conducted by Ling Li, et al. to acquire $CoSe_2$-doped ZnSe/rGO nanocomposite electrode materials, with rGO serving as the substrate for growth. A 3-electrodes test was conducted with a current-density (E_d) of 1.0 A/g may have a specific-capacity of up to 2075.43 F/g, and after 10,000 cycles, the capacity retention rate is 83.19%. At an energy-density of 85.33 W/kg, the constructed GCZS-2//AC (asymmetric capacitors) has a high-power-density of 1279.95 W/kg [51]. Mahalakshmi. K, et. al was prepared the $CoSe_2$ by hydrothermal route and the $CoSe_2$ was distributed evenly around the surface of rGO. In a 3-M LiOH electrolyte, the rGO-incorporated $CoSe_2$ (rGO@CoSe) nanocomposite's electrochemical characteristics show a specific-capacitance of 202 F/g at a scan rate of 5 mV/s. Because of the conductivity of rGO, the produced nanocomposite exhibits optimal capacitor behaviour and a lower charge transfer resistance between the electrode and electrolyte [52]. Furthermore $CoSe_2$@rGO nano-hybrid shows the good specific capacitance was found to 449 F/g at 2 mV/s and 219 F/g at 0.5 A/ g, cyclic stability is 91.3%, and capacitance retention for 5000 cycles, relative to pure $CoSe_2$ specific-capacitance is found to 196 F/g at 2 mV/s, 99 F/g at 0.5 A/g is shown in Fig. 5 [53]. The $NiSe_2$-CoSe, self-supported-nanoflowers (NFs) have a high-specific capacitance value of 672 F/g at a current-density of 1 A/g and 89.6 % capacity retention after 20,000 cycles at 20 A/g. A cheap wet-chemical method was used to make the $NiSe_2$-CoSe hybrid, which had an amazing energy-density of 54.9 Wh/kg at a high power-density of 8460 W/kg with 89.6% preservation after 20,000 cycles, $NiSe_2$-CoSe is very stable [54]. The high specific capacitance of 1521.6 F/g at 1 A/g was shown the CuCoSe-doped rGO-NF electrode, and it has enhanced rate performance and a remarkable capacitance retention of 99.5% above 6000 cycles. Once put together, the ASC device has a remarkable energy-density of 19.8 Wh/kg at a power-density of 750 W/kg. Additionally, it exhibits an exceptional capacitance retention of 86.2% even after undergoing 6000 cycles [55]. The $CoSe_2$/C electrodes have demonstrated remarkable electrochemical performance due to their unique composition. Their specific capacitance is found to 462 F/g when the current-density is 5 A/g. They are able to retain 100% of their capacitance at a current-density of 10 A/g for 10,000 cycles. Simultaneously, an asymmetrical SCs (//ASC) constructed, demonstrating an energy-density of 20.6 Wh/kg, together with a power-density of 698.8 W/kg, and exceptional cycle stability [56].

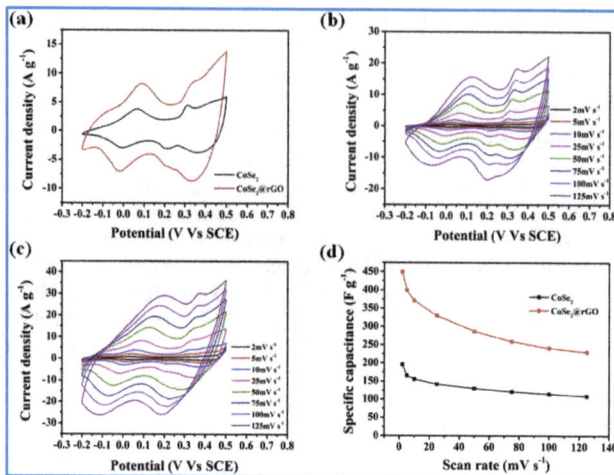

Fig. 5. (a) The CV curves of CoSe$_2$ and 5%-rGO@CoSe$_2$ at 25 mV s^{-1}. (b) CV curves of CoSe$_2$ at various scan rates (c) CV curves of 5%-rGO@CoSe$_2$ at various scan rates. (d) C$_{sp}$ of CoSe$_2$ and 5%-rGO@CoSe electrodes as a function of various scan rates such as (2, 5, 10, 25, 50, 75, 100 and 125 mVs^{-1} respectively).

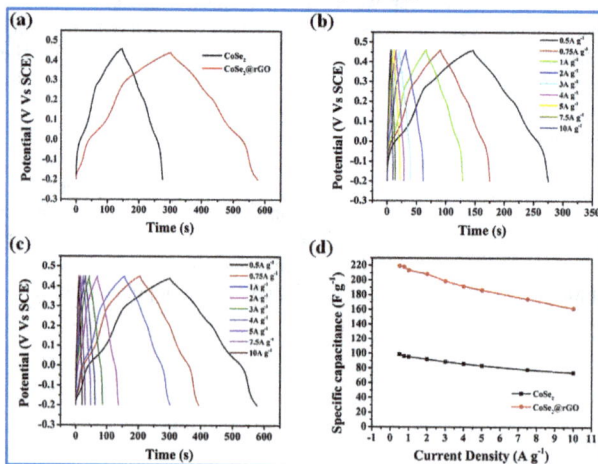

Fig. 6 (a) GCD of CoSe$_2$ and 5%-rGO@CoSe$_2$ nano-hybrid at a constant current-density of 0.5 A /g. (b) GCD measurement of CoSe$_2$ at various current densities. (c) GCD studies of 5%-rGO@CoSe$_2$ at various current densities. (d) C$_{sp}$ of CoSe$_2$ and 5%-rGO@CoSe$_2$ electrodes as a function of various current-densities.

4.2   Graphene-NiSe/rGO//AC composites based supercapacitors

The researchers developed a new nanocomposite called $NiSe_2@rGO$, which consists of reduced graphene oxide (rGO) coated with $NiSe_2$ nanoparticles. This nanocomposite has a high specific-capacity (C_{sp}) of 467 C/g at 1 A/g and demonstrates excellent rate performance, retaining around 72.8% of its capacity at 20 A/g. In addition, a hybrid SCs (HSC) made composed of a positive electrode consisting of $NiSe_2@rGO$ and a negative electrode made of rGO achieves a high energy density of 34.3 Wh/kg and an impressive cycling longevity with around 93% capacity retention after 6500 cycles [57]. The combination of nickel compounds have shown high capacitance and cobalt compounds high cyclic stability and enhances the performance of electrode material based on nickel and cobalt. For instance, Du et al. successfully synthesized $(Ni_xCo_{1-x})_{0.85}Se$ nanosheets on nickel foam by hydrothermal and cation exchange methods, $(Ni_{0.5}Co_{0.5})_{0.85}Se$ nanosheet arrays deliver a higher specific-capacity (C_{sp}) of 430.87 mAh/g at 1 A/g than conventional electrodes. Remarkably, the nanosheets maintained 85.25% of their original composition even after undergoing 3000 cycles [58]. Haichao Chen et al. used a simple one-step solvothermal process to produce $(NiCo)Se_2$. The electrochemical performance of a NiSe–CoSe sample, with a ratio of 4 times NiSe and 2-times CoSe (4:2 ratio) were assessed by constructing hybrid SCs using RGO. These hybrid SCs exhibit both high power and energy densities, with a value of 41.8 Wh/kg at 750 W/kg and 20.3 Wh/kg at 30 kW/kg. After undergoing 1000 cycles, the NiSe-CoSe sample, with a NiSe and CoSe ratio of 4:2, exhibits a specific-capacity (C_{sp}) of 426 C/g at 7 A/g. This specific-capacity (C_{sp}) is significantly higher than that of samples with different NiSe and CoSe ratios, thus indicating its exceptional performance [59]. The wet chemical approach was used to successfully create a $ZnS-CoSe_2$ (ZC-2) nanocomposite for use in hybrid SCs. The capacitance performance of the $ZnS50\%-CoSe_2$ nanocomposite electrode is much better (953.3 F/g) compared to that of pure ZnS (316.6 F/g). The hybrid SCs system achieved an energy density of 46.71 Wh/kg and a power-density (P_d) of 5400 W/kg [60]. Basit Ali Khan, et. al reported the prepared $NiSe_2$-16% rGO electrode has shown an enhance specific-capacitance is 1845.5 F/g at 1.5 A/g, Due to their unique composition, the $CoSe_2/C$ electrodes have demonstrated remarkable electrochemical performance. Their specific capacitance is 462 F/g when the current density is 5 A/g. They are able to retain 100% of their capacitance at a current density of 10 A/g for 10,000 cycles [61]. Moreover, the $Ni_{0.85}Se@C/rGO//AC$ SCs device shows an excellent specific-capacitance 430.87 mAh /g at 1 A/ g· and specific-energy of 45.6 Wh/kg and 31.4 Wh/kg at specific-power of 463.5 W/ kg and 9.4 kW/kg reported by Zhiying Sun and co-workers [62]. The $NiZrSe_3/rGO$ composite was prepared through facile one-pot hydrothermal technique (HT). The obtained specific-capacity (C_{sp}) of $NiZrSe_3/rGO$ is 2750.8 F/g which is more than 1415.7 F/g achieved by $NiZrSe_3$ at 2 A/g. The integration of rGO reduced the agglomeration of $NiZrSe_3$ nanoparticles. The b-Fit value of $NiZrSe_3/rGO$ was 0.61, have shown mixed charge storage mechanism. $NiZrSe_3/rGO$ was employed as a faradic type electrode material for making hybrid asymmetric supercapacitors (ASCs). $NiZrSe_3/rGO//Act-C$ ASC reached an amazing energy-density of 53.22 Wh/kg at power-density (P_d) of 990 W/kg [63].

4.3   Graphene-MnSe/rGO//AC composites based supercapacitors

The $MnSe_2/rGO$ exhibited an exceptional specific capacitance of 326 F/g/145 C/g at a current density of 1.5 A/g, along with an outstanding longevity of 99.94% after 10,000 cycles. Additionally, it demonstrated fast charge transport responses during electrochemical processes. In addition, an HSC (hybrid supercapacitor) device was fabricated using a composite of $MnSe_2/rGO$

(manganese selenide/reduced graphene oxide). The positive and negative electrodes of this device were made of phosphine-based covalent organic frameworks, specifically denoted as Phos-COF-1. This MnSe$_2$/rGO//Phos-COF-1 HSC device demonstrated the ability to increase the voltage to 1.6 V, providing sufficient energy to power a yellow LED with a high level of brightness. The specific energy of this device was measured at 16.6 Wh/kg/800 W/kg, and it maintained a specific energy of 10.0 Wh/kg even when subjected to a high specific power of 7200 W/kg [64]. The hybrid of α-MnS and N-rGO shown excellent electrochemical performance. The hybrid material used as an electrode in supercapacitors exhibits a specific capacitance of 933.6 F/g at a current density of 1 A/g. Even at a high discharge current density of 20 A/g, the specific capacitance maintains at 469.1 F/g [65]. The Javaid Ismail, et. al reported the 5%GO composite containing MnSe demonstrated a capacitance of 180 F/g, along with the lowest charge transfer resistance. The created and built an asymmetric supercapacitor (ASC) with a capacitance of 56.25 F/g, an energy density of 31.25 Wh/kg, and a power-density (P$_d$) of 6779.20 W/kg, focusing on practical considerations. Surprisingly, our artificially created ASC composed of MnSe-GO-//AC exhibited exceptional cycling performance, maintaining 86.3% of its original capacitance after undergoing 5000 consecutive charge-discharge cycles [66]. Moreover, the manganese selenide (MnSe), which has a high theoretical capacity and reversibility, is a very promising material for the anode of sodium-ion batteries. The sodium storage by MnSe-NC is regulated by surface pseudocapacitance, showing a promising rate capability. As-prepared MnSe-NC exhibits promising rate performance specific-capacity (C$_{sp}$) 237 mAh/g at 2 A/g, and long-term cycling stability (232 mAh/g after 2000 cycles) for sodium storage [67]. A complete sodium-ions cell was studied using a MnSe-NC anode paired with a Na$_3$V$_2$(PO$_4$)$_2$@rGO cathode. The anode showed a specific-capacity of 195 mAh/g at a current-density of 0.1 A/g [68].

Table. 2. Listed the comparison of some GO/rGO-doped metal chalcogenides composites electrode materials and their performance in specific capacitance.

S.No.	Electrode Materials	Method	Charge discharge cycle	Specific Capacitance	Ref.
1	CoSe$_2$-doped ZnSe/rGO	Co-precipitation	10000	2075.43 Fg^{-1} at 1.0 Ag^{-1}	[51]
2	CoSe$_2$	Hydrothermal	5000	202 Fg^{-1} at 5 mVs^{-1}	[52]
3	CoSe@rGO and Pure CoSe	Hydrothermal	5000	449 Fg^{-1} at 2 mVs^{-1} 196 Fg^{-1} at 2 mVs^{-1}	[53]
4	NiSe$_2$-CoSe	Wet-Chemical	20000	672 Fg^{-1} at 1 Ag^{-1}	[54]
5	CuCoSe@rGO-NF	co-precipitation	6000	1521.6 Fg^{-1} at 1 Ag^{-1}	[55]
6	CoSe/C	co-precipitation	10000	462 F g^{-1} at 5 A g^{-1}	[56]
7	NiSe$_2$@rGO	Hydrothermal	6500	467 C g^{-1} at 1 A g^{-1}	[57]
8	(Ni$_x$Co$_{1-x}$)$_{0.85}$Se	Cation-exchange	3000	430.87 mAh g^{-1} at 1 A g^{-1}	[58]
9	(NiCo)Se$_2$	One-pot solvothermal	1000	426 Cg^{-1} at 7 Ag^{-1}	[59]
10	ZnSe 50% CoSe$_2$	Co-precipitation	5000	953.3 Fg^{-1} at 1 Ag^{-1}	[60]

11	NiSe$_2$ 16% rGO	Hydrothermal	5000	1845 F g^{-1} at 1.5 A g^{-1}	[63]
12	MnSe/rGO	Hydrothermal	10000	326 F g^{-1} at 1.5 Ag^{-1}	[64]
13	5%GO@MnSe	Hydrothermal	5000	180 F g^{-1} at 1Ag^{-1}	[65]
14	MnSe-NC	Hydrothermal	2000	237 mAh g^{-1} at 2 A g^{-1}	[68]

Summary and overviews.

In this chapter, we discussed the growing need for energy and why it is critical to find solutions by creating energy storage technologies. We also covered the basics of supercapacitors, including their remarkable storage capacity and the potential good they might bring to the future world. We discussed the potential of graphene and reduced graphene oxide (rGO) doped metal chalcogenides based symmetric and asymmetric supercapacitors have shown enhance electrical conductivity and substantial pseudo-capacitance. We examined the techniques for synthesizing and preparing electrodes made by metal chalcogenides, such as GO/rGO-doped metal chalcogenides (CoSe, NiSe, and MnSe) composites. We also explored GO and rGO-doped metal chalcogenide composites for supercapacitor applications. Finally, we covered the methods used to generate GO/rGO-doped metal chalcogenides composites and how well these performed as electrodes.

References

[1] Trudeau, N. and I. Murray (2011), "Development of Energy Efficiency Indicators in Russia", IEA Energy Papers, No. 2011/01, OECD Publishing. https://doi.org/10.1787/5kgk7w8v4dhl-en

[2] OECD (2011a), Towards Green Growth: Monitoring Progress: OECD Indicators, OECD Green Growth Studies, OECD Publishing. https://doi.org/10.1787/9789264111356-en

[3] IEA (2023), World Energy Outlook 2023, IEA, Paris https://www.iea.org/reports/world-energy-outlook-2023, Licence: CC BY 4.0 (report); CC BY NC SA 4.0 (Annex A).

[4] Haščič, I., N. Johnstone, F. Watson, and C. Kaminker (2010), "Climate Policy and Technological Innovation and Transfer: An Overview of Trends and Recent Empirical Results", OECD Environment Working Papers, No. 30, OECD Publishing. https://doi.org/10.1787/5km33bnggcd0-en

[5] IEA (2009a), "Towards a More Energy Efficient Future: Applying Indicators to Enhance Energy Policy", OECD/IEA, Paris, available at: www.iea.org/papers/2009/indicators_brochure2009.pdf

[6] Mohtasham, J. (2015). Review Article-Renewable Energies. Energy Procedia, 74, 1289–1297. https://doi.org/10.1016/j.egypro.2015.07.774

[7] Neha, & Joon, R. (2021). Renewable Energy Sources: A Review. Journal of Physics: Conference Series, 1979(1), 012023. https://doi.org/10.1088/1742-6596/1979/1/012023

[8] Erdiwansyah, Mahidin, Husin, H. et al. A critical review of the integration of renewable energy sources with various technologies. Prot Control Mod Power Syst **6**, 3 (2021). https://doi.org/10.1186/s41601-021-00181-3

[9] C. Allen, G. Metternicht, T. Wiedmann, National pathways to the Sustainable Development Goals (SDGs): a comparative review of scenario modelling tools, Environ. Sci. Policy 66 (2016) 129–207. https://doi.org/10.1016/j.envsci.2016.09. 008

[10]   D.L. McCollum, et al., Connecting the Sustainable Development Goals by Their Energy Inter-linkages, International Institute for Applied System Analysis (IIASA), Laxenburg, 2017. http://pure.iiasa.ac.at/14567/1/WP-17-006

[11]   Gielen, D., Boshell, F., Saygin, D., Bazilian, M. D., Wagner, N., & Gorini, R. (2019). The role of renewable energy in the global energy transformation. Energy Strategy Reviews, 24, 38–50. https://doi.org/10.1016/j.esr.2019.01.006

[12]   F.F. Nerini, et al., Mapping synergies and trade-offs between energy and the sustainable development goals, Nat. Energy 1 (2017) 2058–7546. https://doi.org/10. 1038/s41560-017-0036-5

[13]   Glavin, M. E., & Hurley, W. G. (2012). Optimisation of a photovoltaic battery ultracapacitor hybrid energy storage system. Solar Energy, 86(10), 3009–3020. https://doi.org/10.1016/j.solener.2012.07.005

[14]   Schaeck, S., Stoermer, A. O., Albers, J., Weirather-Koestner, D., & Kabza, H. (2011). Lead-acid batteries in micro-hybrid applications. Part II. Test proposal. Journal of Power Sources, 196(3), 1555–1560. https://doi.org/10.1016/j.jpowsour.2010.08.07

[15]   Chen, H., Cong, T. N., Yang, W., Tan, C., Li, Y., & Ding, Y. (2009). Progress in electrical energy storage system: A critical review. Progress in Natural Science, 19(3), 291–312. https://doi.org/10.1016/j.pnsc.2008.07.014

[16]   Omar, N., Van Mierlo, J., Verbrugge, B., & Van den Bossche, P. (2010). Power and life enhancement of battery-electrical double layer capacitor for hybrid electric and charge-depleting plug-in vehicle applications. Electrochimica Acta, 55(25), 7524–7531. https://doi.org/10.1016/j.electacta.2010.03.039

[17]   Omar, N., Van Mierlo, J., Van Mulders, F., & Van den Bossche, P. (2009). Assessment of Behaviour of Super Capacitor-battery System in Heavy Hybrid Lift Truck Vehicles. Journal of Asian Electric Vehicles, 7(2), 1277–1282. https://doi.org/10.4130/jaev.7.1277

[18]   Abdel Maksoud, M.I.A., Fahim, R.A., Shalan, A.E. et al. Advanced materials and technologies for supercapacitors used in energy conversion and storage: a review. Environ Chem Lett **19**, 375–439 (2021). https://doi.org/10.1007/s10311-020-01075-w

[19]   Ho, J., Jow, T. R., & Boggs, S. (2010). Historical introduction to capacitor technology. IEEE Electrical Insulation Magazine, 26(1), 20-25. https://doi.org/10.1109/mei.2010.5383924

[20]   Şahin, M., Blaabjerg, F., & Sangwongwanich, A. (2022). A Comprehensive Review on Supercapacitor Applications and Developments. Energies, 15(3), 674. https://doi.org/10.3390/en15030674

[21]   Olabi, A. G., Abbas, Q., al Makky, A., & Abdelkareem, M. A. (2022). Supercapacitors as next generation energy storage devices: Properties and applications. Energy, 248, 123617. https://doi.org/10.1016/j.energy.2022.123617

[22]   Cigolotti, V., Genovese, M., Piraino, F., & Fragiacomo, P. (2023). Applications –
Stationary | Stationary Energy Storage System: Overview. In Reference Module in
Chemistry, Molecular Sciences and Chemical Engineering. Elsevier.
https://doi.org/10.1016/B978-0-323-96022-9.00091-8

[23]   Lemian, D., & Bode, F. (2022). Battery-Supercapacitor Energy Storage Systems for
Electrical Vehicles: A Review. Energies, 15(15), 5683. https://doi.org/10.3390/en15155683

[24]   M. K. Andreev, "An Overview of Supercapacitors as New Power Sources in Hybrid
Energy Storage Systems for Electric Vehicles," *2020 XI National Conference with
International Participation (ELECTRONICA)*, Sofia, Bulgaria, 2020, pp. 1-4.
https://doi.org/10.1109/ELECTRONICA50406.2020.9305104

[25]   Yadlapalli, R. T., Alla, R. R., Kandipati, R., & Kotapati, A. (2022). Super capacitors for
energy storage: Progress, applications and challenges. Journal of Energy Storage, 49,
104194. https://doi.org/10.1016/j.est.2022.104194

[26]   Stevic, Z., & Radovanovic, I. (2023). Supercapacitors: The Innovation of Energy Storage.
In Updates on Supercapacitors. IntechOpen. https://doi.org/10.5772/intechopen.106705

[27]   Gromadskyi, D.G., Hromadska, L.I. Bivariant mechanical tuning of porous carbon
electrodes for high-power and high-energy supercapacitors. Surf. Engin.
Appl.Electrochem. **52**, 584–593 (2016). https://doi.org/10.3103/S1068375516060077

[28]   Pandolfo, A. G., & Hollenkamp, A. F. (2006). Carbon properties and their role in
supercapacitors. Journal of Power Sources, 157(1), 11–27.
https://doi.org/doi:10.1016/j.jpowsour.2006.02.065

[29]   Taer, E., Ridholana, R. E., Apriwandi, Taslim, R., & Agustino. (2021). Effective cost and
high-performance supercapacitor electrodes from Syzygium oleana leave biomass wastes.
Journal of Physics: Conference Series, 1811(1), 012134. https://doi.org/10.1088/1742-
6596/1811/1/012134

[30]   Mani, Alagiri, Kamali, Khosro Zangeneh, Pandikumar, Alagarsamy, Lim, Yee Seng,
Lim, Hong Ngee, Huang, Nay Ming, pp. 225-244, Graphene-Polypyrrole Nanocomposite:
An Ideal Electroactive Material for High Performance Supercapacitors.
https://doi.org/10.1002/9781119131816.ch7

[31]   Wang, S., Wei, T., Qi, Z. (2008). Supercapacitor Energy Storage Technology and its
Application in Renewable Energy Power Generation System. In: Goswami, D.Y., Zhao, Y.
(eds) Proceedings of ISES World Congress 2007 (Vol. I – Vol. V). Springer, Berlin,
Heidelberg. https://doi.org/10.1007/978-3-540-75997-3_566.

[32]   Zhang, J., Gu, M., & Chen, X. (2023). Supercapacitors for renewable energy
applications: A review. Micro and Nano Engineering, 21, 100229.
https://doi.org/doi.org/10.1016/j.mne.2023.100229

[33]   A. Burke, "Ultracapacitors: Why, How, and Where Is the Technology," Journal of Power
Sources, Vol. 91, No. 1, 2000, pp. 37-50. https://doi.org/10.1016/S0378-7753(00)00485-7

[34]   Yadlapalli, R. T., Alla, R. R., Kandipati, R., & Kotapati, A. (2022). Super capacitors for
energy storage: Progress, applications and challenges. Journal of Energy Storage, 49,
104194. https://doi.org/10.1016/j.est.2022.104194

[35]   Kurra, N., & Jiang, Q. (2022). Supercapacitors. In Storing Energy (pp. 383–417).
Elsevier. https://doi.org/10.1016/B978-0-12-824510-1.00017-9

[36] Bilal, M., Rehman, Z. U., Hou, J., Ali, S., Ullah, S., & Ahmad, J. (2022). Metal oxide–carbon composite: synthesis and properties by using conventional enabling technologies. In Metal Oxide-Carbon Hybrid Materials (pp. 25–60). Elsevier. https://doi.org/10.1016/B978-0-12-822694-0.00021-1

[37] Imran, M., Qasam, K., Safdar, S. et al. Hydrothermally synthesized CuNiS@CNTs composite electrode material for hybrid supercapacitors and non-enzymatic electrochemical glucose sensor. J Mater Sci: Mater Electron 35, 441 (2024). https://doi.org/10.1007/s10854-024-12197-0

[38] Forouzandeh, P., Kumaravel, V., & Pillai, S. C. (2020). Electrode Materials for Supercapacitors: A Review of Recent Advances. Catalysts, 10(9), 969. https://doi.org/10.3390/catal10090969

[39] Ali, Z., Yaqoob, S., Yu, J., & D'Amore, A. (2024). Critical review on the characterization, preparation, and enhanced mechanical, thermal, and electrical properties of carbon nanotubes and their hybrid filler polymer composites for various applications. Composites Part C: Open Access, 13, 100434. https://doi.org/10.1016/j.jcomc.2024.100434

[40] Premanand, G., Sridevi, D. V., Perumal, S., Maiyalagan, T., Rodney, J. D., & Ramesh, V. (2022). New hybrid semiconducting CdSe and Fe doped CdSe quantum dots based electrochemical capacitors. Materials Science and Engineering: B, 286, 116015. https://doi.org/10.1016/j.mseb.2022.116015

[41] A, A., V, R., Perumal, S., & Nayak, P. K. (2023). Structural, vibrational and electrochemical studies of bulk and nano SnSb for supercapacitor application. Journal of Alloys and Compounds, 969, 172293. https://doi.org/10.1016/j.jallcom.2023.172293

[42] Abhishek A, Ravi K. R, Tushar H. Rana, Rajasekar Parasuraman, Suresh Perumal, Ramesh V, Thermoelectric, mechanical and electrochemical properties of pure single-phase FeSb, Ceramics International, 2024, ISSN 0272-8842. https://doi.org/10.1016/j.ceramint.2024.04.403.

[43] Macías-García, A., Torrejón-Martín, D., Díaz-Díez, M. Á., & Carrasco-Amador, J. P. (2019). Study of the influence of particle size of activate carbon for the manufacture of electrodes for supercapacitors. Journal of Energy Storage, 25, 100829. https://doi.org/10.1016/j.est.2019.100829

[44] Dujearic-Stephane, K., Gupta, M., Kumar, A., Sharma, V., Pandit, S., Bocchetta, P., & Kumar, Y. (2021). The Effect of Modifications of Activated Carbon Materials on the Capacitive Performance: Surface, Microstructure, and Wettability. Journal of Composites Science, 5(3), 66. https://doi.org/10.3390/jcs5030066

[45] Ivanovskii, A. L. (2012). Graphene-based and graphene-like materials. Russian Chemical Reviews, 81(7), 571–605. https://doi.org/10.1070/RC2012v081n07ABEH004302

[46] Li, X., Zhu, Y., Cai, W., Borysiak, M., Han, B., Chen, D., Piner, R. D., Colombo, L., & Ruoff, R. S. (2009). Transfer of Large-Area Graphene Films for High-Performance Transparent Conductive Electrodes. Nano Letters, 9(12), 4359–4363. https://doi.org/10.1021/nl902623y

[47] Tîlmaciu, C.-M., & Morris, M. C. (2015). Carbon nanotube biosensors. Frontiers in Chemistry, 3. https://doi.org/10.3389/fchem.2015.00059

[48] Mohan, V. B., Lau, K., Hui, D., & Bhattacharyya, D. (2018). Graphene-based materials and their composites: A review on production, applications and product limitations.

Composites Part B: Engineering, 142, 200–220.
https://doi.org/10.1016/j.compositesb.2018.01.013

[49]   Gandhi, M. R., Vasudevan, S., Shibayama, A., & Yamada, M. (2016). Graphene and Graphene-Based Composites: A Rising Star in Water Purification - A Comprehensive Overview. ChemistrySelect, 1(15), 4358–4385. https://doi.org/10.1002/slct.201600693

[50]   Pandey, A., & Chauhan, P. (2023). Functionalized graphene nanomaterials: Next-generation nanomedicine. In Functionalized Carbon Nanomaterials for Theranostic Applications (pp. 3–18). Elsevier. https://doi.org/10.1016/B978-0-12-824366-4.00020-0

[51]   Li, L., Liu, W., Wei, Z., & Li, Z. (2024). GO induces ZIF-8 to produce hollow nanobox $CoSe_2$@ZnSe/rGO for high-performance supercapacitors. Materials Today Sustainability, 26, 100775. https://doi.org/10.1016/j.mtsust.2024.100775

[52]   Mahalakshmi, K., Jenila, R. M., Vivek, E., Potheher, I. V., & Thangaraj, V. (2023). Studies on dielectric, thermal and electrochemical characteristics of rGO incorporated CoSe2 nanocomposite for energy storage application. Synthetic Metals, 299, 117450. https://doi.org/10.1016/j.synthmet.2023.117450

[53]   Song, Y., Ran, A., Peng, Z., Huang, W., Huang, B., Jian, X., & Mu, C. (2019). Cobalt Diselenide@Reduced graphene oxide based nanohybrid for supercapacitor applications. Composites Part B: Engineering, 174, 107001. https://doi.org/10.1016/j.compositesb.2019.107001

[54]   Shah, M. S. U., Zuo, X., Shah, A., Al-Saeedi, S. I., Shah, M. Z. U., Alabbad, E. A., Hou, H., Ahmad, S. A., Arif, M., Sajjad, M., & Haq, T. U. (2023). CoSe nanoparticles supported NiSe2 nanoflowers cathode with improved energy storage performance for advanced hybrid supercapacitors. Journal of Energy Storage, 65, 107267. https://doi.org/10.1016/j.est.2023.107267

[55]   Tavakoli, F., Rezaei, B., Taghipour Jahromi, A. R., & Ensafi, A. A. (2020). Facile Synthesis of Yolk-Shelled CuCo $_2$ Se $_4$ Microspheres as a Novel Electrode Material for Supercapacitor Application. ACS Applied Materials & Interfaces, 12(1), 418–427. https://doi.org/10.1021/acsami.9b12805

[56]   Wang, H., Shu, T., Yuan, J., Li, Y., Lin, B., Wei, F., Qi, J., & Sui, Y. (2022). Highly stable lamellar array composed of CoSe2 nanoparticles for supercapacitors. Colloids and Surfaces A: Physicochemical and Engineering Aspects, 633, 127789. https://doi.org/10.1016/j.colsurfa.2021.127789

[57]   Gu, Y., Fan, L.-Q., Huang, J.-L., Geng, C.-L., Lin, J.-M., Huang, M.-L., Huang, Y.-F., & Wu, J.-H. (2019). N-doped reduced graphene oxide decorated NiSe2 nanoparticles for high-performance asymmetric supercapacitors. Journal of Power Sources, 425, 60–68. https://doi.org/10.1016/j.jpowsour.2019.03.123

[58]   Wang, J., Sarwar, S., Song, J., Du, L., Li, T., Zhang, Y., Li, B., Guo, Q., Luo, J., & Zhang, X. (2022). One-step microwave synthesis of self-supported CoSe2@NiSe2 nanoflowers on 3D nickel foam for high performance supercapacitors. Journal of Alloys and Compounds, 892, 162079. https://doi.org/10.1016/j.jallcom.2021.162079

[59]   Chen, H., Fan, M., Li, C., Tian, G., Lv, C., Chen, D., Shu, K., & Jiang, J. (2016). One-pot synthesis of hollow NiSe–CoSe nanoparticles with improved performance for hybrid supercapacitors. Journal of Power Sources, 329, 314–322. https://doi.org/10.1016/j.jpowsour.2016.08.097

[60]   Ahmad, S. A., Shah, M. Z. U., Hussain, I., Arif, M., Song, P., Al-Saeedi, S. I., Sajjad, M., Ahmad, I., Aftab, J., Huang, T., & Shah, A. (2023). CoSe2@ZnS microsphere arrays with remarkable electrochemical performance for hybrid asymmetric supercapacitor. Journal of Energy Storage, 73, 109090. https://doi.org/10.1016/j.est.2023.109090

[61]   Khan, B. A., Hussain, R., Shah, A., Mahmood, A., Shah, M. Z. U., Ismail, J., Rahman, S. ur, Sajjad, M., Assiri, M. A., Imran, M., & Javed, M. S. (2022). NiSe2 nanocrystals intercalated rGO sheets as a high-performance asymmetric supercapacitor electrode. Ceramics International, 48(4), 5509–5517. https://doi.org/10.1016/j.ceramint.2021.11.095

[62]   Sun, Z., Liu, F., Wang, J., Hu, Y., Fan, Y., Yan, S., Yang, J., & Xu, L. (2020). Tiny Ni0.85Se nanosheets modified by amorphous carbon and rGO with enhanced electrochemical performance toward hybrid supercapacitors. Journal of Energy Storage, 29, 101348. https://doi.org/10.1016/j.est.2020.101348

[63]   Fu, H., Chen, Y., Ren, Z., Xiao, Y., Liu, Y., Zhang, X., & Tian, G. (2018). Highly dispersed of Ni0.85Se nanoparticles on nitrogen-doped graphene oxide as efficient and durable electrocatalyst for hydrogen evolution reaction. Electrochimica Acta, 262, 107–114. https://doi.org/10.1016/j.electacta.2017.12.144

[64]   Peng, H., Wei, C., Wang, K., Meng, T., Ma, G., Lei, Z., & Gong, X. (2017). Ni 0.85 Se@MoSe 2 Nanosheet Arrays as the Electrode for High-Performance Supercapacitors. ACS Applied Materials & Interfaces, 9(20), 17067–17075. https://doi.org/10.1021/acsami.7b02776

[65]   Sajjad, M., Ismail, J., Shah, A., Mahmood, A., Shah, M. Z. U., Rahman, S. ur, & Lu, W. (2021). Fabrication of 1.6V hybrid supercapacitor developed using MnSe2/rGO positive electrode and phosphine based covalent organic frameworks as a negative electrode enables superb stability up to 28,000 cycles. Journal of Energy Storage, 44, 103318. https://doi.org/10.1016/j.est.2021.103318

[66]   Quan, H., Cheng, B., Chen, D., Su, X., Xiao, Y., & Lei, S. (2016). One-pot synthesis of α-MnS/nitrogen-doped reduced graphene oxide hybrid for high-performance asymmetric supercapacitors. Electrochimica Acta, 210, 557–566. https://doi.org/10.1016/j.electacta.2016.05.031

[67]   Ismail, J., Sajjad, M., Ullah Shah, M. Z., Hussain, R., Khan, B. A., Maryam, R., Shah, A., Mahmood, A., & Rahman, S. ur. (2022). Comparative capacitive performance of MnSe encapsulated GO based nanocomposites for advanced electrochemical capacitor with rapid charge transport channels. Materials Chemistry and Physics, 284, 126059. https://doi.org/10.1016/j.matchemphys.2022.126059

[68]   Hu, L., He, L., Wang, X., Shang, C., & Zhou, G. (2020). MnSe embedded in carbon nanofibers as advanced anode material for sodium ion batteries. Nanotechnology, 31(33), 335402. https://doi.org/10.1088/1361-6528/ab8e78

Emerging Materials for Next Frontier Energy and Environment Applications   Materials Research Forum LLC
Materials Research Foundations 170 (2024) 21-40   https://doi.org/10.21741/9781644903292-2

Chapter 2

2D Materials (2DM) based photocatalyst for environmental remediation

R. Roshan Chandrapal, G. Bakiyaraj[*]

Department of Physics and Nanotechnology, Faculty of Engineering and Technology, SRM Institute of Science and Technology, Kattankulathur, 603203, Tamil Nadu, India

[*]bakiyarg@srmist.edu.in

Abstract

Creating a sustainable environment is one of the most important issues facing modern society. Together with an increase in human activity and a rise in per capita income, there has been a noticeable population boom. A growing number of people are becoming concerned about whether the natural resources of our planet will be viable in the long run. In addition, pollution of the air, water, and soil has resulted from industrialization, urbanization, and contemporary farming methods. Given the current situation, widespread exploitation and the widespread presence of hazardous substances are posing growing challenges to the ability of future generations to flourish. Water is separated into hydrogen and oxygen in this process, with the hydrogen being used as fuel. For semiconductor-based photocatalysts and materials to be used effectively in photocatalytic degradation processes for environmental remediation and energy conversion applications, a thorough understanding of their characteristics and behavior is essential. 2DM are an emerging class of nanomaterials with exceptional properties. These composites combine the unique qualities of specific 2DM to create new capabilities. To achieve an increased efficiency, a variety of 2D composites (0D/2D, 1D/2D, 2D/2D, and 3D/2D) are used. Two unique 2D materials that can be stacked to create designer materials with particular properties are graphene and hexagonal boron nitride. Materials derived from 2DM-composite provide exciting prospects in the realm of photocatalysts for the degradation of organic pollutants. Because of their varied uses and customizability, they are at the forefront of developing advanced materials for a range of applications. More groundbreaking 2DM composite with game-changing implications for photocatalysis could be in store.

Keywords

Semiconductor Photocatalyst, 2D Materials, Tunable Bandgap, CO_2 Reduction, Pollutant Degradation, Water Splitting

Emerging Materials for Next Frontier Energy and Environment Applications   Materials Research Forum LLC
Materials Research Foundations 170 (2024) 21-40   https://doi.org/10.21741/9781644903292-2

Contents

2D Materials (2DM) based photocatalyst for environmental remediation......................**21**

1.   Introduction...**22**

2.   Properties of 2DM..**25**

2.1   Unique atomic structure of 2DM...25

2.2   Tunability of electronic bandgaps 26

2.3   Surface area and its advantage...26

2.4   Conductivity, stability, and charge separation efficiency in 2DM 27

3.   Applications of 2DM in Photocatalysis **28**

3.1   Water splitting 28

3.2   CO$_2$ reduction 30

3.3   Pollutant degradation..32

Conclusion **33**

Reference **34**

1.   Introduction

Harnessing the power of sunlight to trigger chemical reactions, photocatalysis has been widely regarded as a potential solution for tackling environmental challenges. Similar to the world it aims to improve, this field is currently experiencing a significant shift led by a new generation of remarkable individuals. Picture an innovative technology harnessing the immense energy of sunlight to tackle urgent environmental issues and open doors to sustainable energy solutions [1]. Discover the fascinating realm of photocatalysis, where certain substances known as photocatalysts harness the power of light to undergo remarkable transformations. These catalysts harness the power of sunlight to initiate a series of chemical reactions that have the potential to dismantle pollutants, extract water molecules, and transform CO$_2$. Photocatalysts plays a crucial role in ensuring a cleaner environment by effectively decomposing harmful organic compounds, bacteria, and other impurities. Photocatalysts provide a practical solution for the production of clean hydrogen fuel, CO$_2$ reduction and pollution degradation, etc., which is crucial for reducing carbon emissions in transportation and energy generation [2]. Photocatalysis presents a hopeful solution to address these urgent environmental concerns, providing optimism for a more sustainable future. It provides a greener alternative to conventional pollution control methods that depend on harmful chemicals and use significant energy. In addition, photocatalysis is crucial for the advancement of a circular economy. It contributes to the prevention of pollution and encourages the reuse of resources by producing clean fuels and recycling waste materials. However, the journey is just beginning. Despite the progress made in photocatalysis research, there is still a need for additional research and development to address challenges related to material optimization, production scalability, and cost effectiveness [3-5].

Emerging Materials for Next Frontier Energy and Environment Applications   Materials Research Forum LLC
Materials Research Foundations 170 (2024) 21-40   https://doi.org/10.21741/9781644903292-2

The story behind the emergence of 2DM is a captivating journey into the realm of science and technology. Steeped in a realm of unexpected revelations, limitless potentials, and a fierce race to unveil revolutionary applications, the adventure commences in 2004 with Novoselov and Geim's triumphant isolation of graphene a remarkable carbon atom sheet [6]. In this critical moment, there was a wave of interest in 2DM, ignited by the extraordinary properties of graphene, such as its superconductivity, impressive strength, and remarkable flexibility. Scientists were intrigued by this, leading them to conduct a study on non-carbon 2D materials like transition metal dichalcogenides (TMDs) [7], MOF [8], Mxenes [9], etc. They found some interesting features, such as the ability to adjust bandgaps, impressive light absorption, and the potential for valleytronics. The rational design of 2D photocatalyst was shown in Fig. 1 [10].

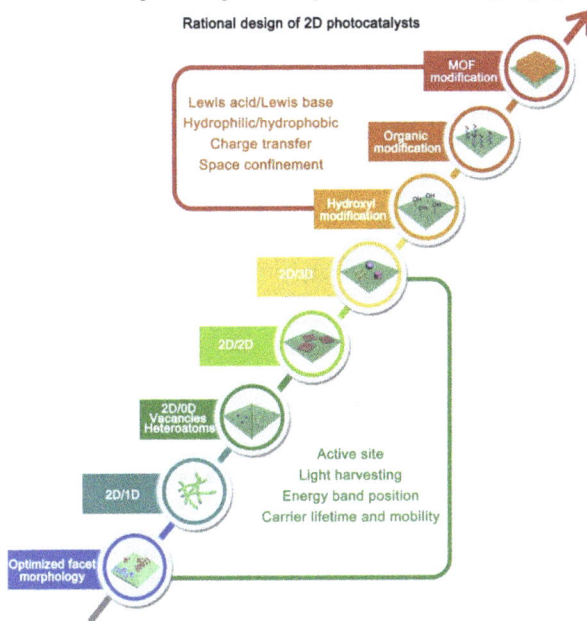

Fig. 1 Rational design of 2D photocatalyst

Every material has distinct properties that contribute to its composition, including quantum confinement, large surface area, and remarkable flexibility. Similar to a physicist, it is evident that the realm of 2DM is still in its nascent phase [11-12].

Understanding the distinct properties of metallic bonding which facilitate the formation of three-dimensional structures through extensive and non-specific bonding poses a significant hurdle in the quest to create stable, atomically-thin two-dimensional metals. These materials showcase distinct properties that set them apart from their larger counterparts, showcasing customizable plasmonic resonances, heightened catalytic activity, and even superconductivity [13].

Emerging Materials for Next Frontier Energy and Environment Applications   Materials Research Forum LLC
Materials Research Foundations 170 (2024) 21-40   https://doi.org/10.21741/9781644903292-2

Scientists employ a method known as bottom-up synthesis to create 2DM. These methods involve the meticulous growth of films, atom by atom, using techniques like chemical vapor deposition or molecular beam epitaxy. With the precision of a scientist, one can achieve complete mastery over thickness and structure [14]. Alternatively, top-down approaches prioritize the elimination of substantial amounts of metals or their application onto substrates such as graphene. Nevertheless, attaining accurate manipulation of size and thickness poses a more substantial obstacle to this approach [15]. Delving into the realm of 2DM reveals exciting opportunities and breakthroughs, broadening our understanding of how materials behave at the atomic scale. The photocatalytic process was represented in Fig. 2 [16, 17].

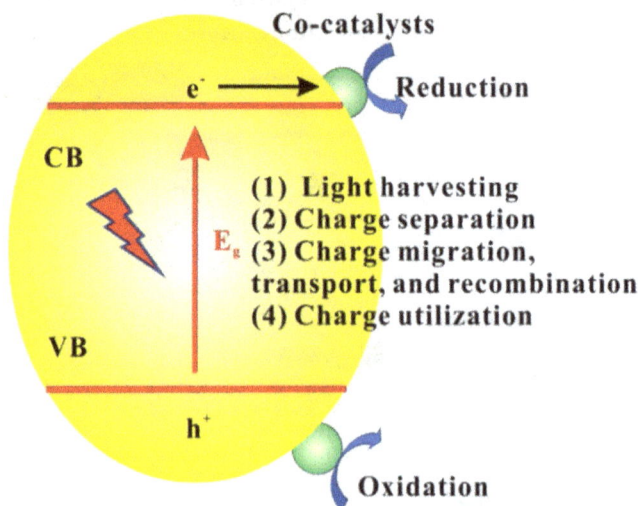

Fig. 2 Photocatalytic process

Their deep understanding of manipulating plasmonic properties gives them a clear advantage. Through meticulous adjustments to the thickness and composition, 2D materials can significantly improve their light absorption capabilities across a broader spectrum of wavelengths. As a result, there is a higher level of light absorption in photocatalytic reactions. Furthermore, the ultra-thin 2D materials demonstrate impressive photocatalytic capabilities thanks to their substantial surface area-to-volume ratio [18]. Examining this property from a scientific perspective reveals a multitude of active sites that significantly improve interactions with reactants and intermediates, resulting in a significant boost in reaction rates. With the unique electronic structure of 2DM, surface activity is further enhanced, providing the opportunity for customization to optimize charge separation and carrier mobility.

Through the application of meticulous optimization techniques, the efficiency of the photocatalytic process is significantly enhanced, leading to a notable improvement in the conversion of light energy into chemical energy [19]. Having a deep understanding of the potential for 2DM to hybridize adds an extra level of excitement to their wide range of applications. Through the combination of different materials, such as semiconductors or metal oxides, it becomes possible to create composites that showcase a unique blend of properties [20]. These hybrid structures harness the distinct properties of each component, leading to enhanced photocatalytic performance and opening up possibilities for groundbreaking applications in energy conversion and environmental remediation.

Investigating the role of 2DM in photocatalysis is a fascinating field of study that pushes the boundaries of our current understanding and showcases their incredible ability to revolutionize light-induced chemical reactions. Just as physicists delve into the characteristics of conventional three-dimensional (3D) materials, the discovery of two-dimensional (2D) materials like graphene and transition metal dichalcogenides (TMDs) has unveiled a plethora of distinctive benefits and properties [21, 22].

Let's explore some benefits of 2DM: Characteristics such as adjustable bandgaps, expansive surface area, distinct electronic properties, the influence of quantum confinement, versatility, and resilience, as well as customized physical and chemical properties. 2D materials such as TMDs have the unique feature of having a modifiable bandgap, which distinguishes them from conventional metals with continuous energy bands. This property is essential for applications in electronics and optoelectronics, where accuracy is paramount.

2.   Properties of 2DM

2.1   Unique atomic structure of 2DM

2DM have a unique atomic structure that sets them apart from their bulk counterparts, leading to fascinating possibilities for various applications. These differences have significant impacts on their properties, and analyzing key aspects reveals the following observations: When it comes to dimensionality, 2DM take the form of a single atomic layer that looks like an extended sheet. By reducing the dimensionality, electron movement is confined to two dimensions, leading to the emergence of unique electronic phenomena [23]. Understanding the bonding and coordination in 2DM is distinct from that of bulk metals, as it relies on less strong interlayer interactions like van der Waals forces or covalent bonds. This property enables more convenient exfoliation and stacking into heterostructures. In 2DM, the reduced coordination number, which refers to the fewer number of neighbors each atom bonds with, has a significant impact on various properties such as conductivity and magnetism. When electrons are confined to a single plane, they experience quantum confinement effects [24]. This causes their energies and behavior to change, resulting in distinct electronic band structures and optical properties. Understanding the behavior of 2Dm requires considering the significant impact of surface effects. The large surface area-to-volume ratio in these materials amplifies the influence of surface properties, making them more prominent than in bulk materials [25]. It is worth mentioning that 2DM exhibit exceptional carrier mobility, even surpassing certain bulk materials. This characteristic makes them highly promising for the development of high-performance electronics. The thinness of these materials and their delicate interlayer interactions make it possible to adjust their properties through various methods such as doping, stacking, and applying external fields. This opens up the exciting possibility of designing materials with precise functionalities [26]. Furthermore, the distinct atomic arrangement of 2DM can lead to the emergence of novel electronic states, such as superconductivity at elevated temperatures, and the occurrence of phenomena like exciton condensation, which are not typically observed in larger quantities of materials. In general, the unique characteristics of 2DM provide an excellent opportunity for exploration and innovation in the field of materials science and technology.

2.2   Tunability of electronic bandgaps

Exploring the fascinating world of 2DM opens up a wealth of opportunities for controlling electronic bandgaps, providing ways to tailor light absorption and catalytic activity. This intriguing interplay arises from the distinct qualities of 2DM, which distinguish them from their bulk counterparts in captivating ways. 2DM have distinct bandgaps due to their reduced dimensionality and quantum confinement, unlike traditional bulk metals with overlapping valence and conduction bands [27].   With its inherent features, a whole world of possibilities opens up for creative manipulation. These include introducing foreign atoms, using mechanical influence to engineer strain, and stacking layers with different bandgaps to create heterostructures. The tunability of this phenomenon has a profound impact, especially when it comes to light absorption. With their precisely engineered bandgaps, 2DM have the ability to absorb particular wavelengths of light [28]. This opens up exciting opportunities for cutting-edge technologies like advanced light-emitting diodes, photodetectors, and solar cells that offer enhanced efficiency and precise manipulation of emitted light. In addition, the intense confinement of light within the thin atomic layer of 2DM amplifies the interaction between light and matter, which is crucial for photocatalysis applications.

Understanding the bandgap of a 2DM is crucial in determining its catalytic activity, going beyond just light absorption. Understanding the intricate relationship between activation energy and bandgap enables us to enhance reaction rates and unlock the power of efficient catalysis. Furthermore, by manipulating the bandgap and electronic states near the surface, it becomes possible to selectively catalyze reactions [29]. This allows for the creation of materials that encourage specific reaction pathways, resulting in improved product selectivity and reduced unwanted side reactions. There are various instances that demonstrate the practical implications of this ability to adjust. As an example, by adding different elements to Molybdenum disulfide (MoS_2), its bandgap can be adjusted, making it suitable for catalyzing hydrogen evolution under different light wavelengths [30]. Just like a physicist, black phosphorus has a tunable bandgap that makes it a promising option for using sunlight to convert CO_2 into valuable hydrocarbon fuels [31]. With the ability to finely manipulate bandgaps in 2DM, a wide range of opportunities arise, bringing forth a fresh era of customized functionalities in various fields such as electronics, catalysis, and energy conversion.

2.3   Surface area and its advantage

2DM are at the forefront of surface area research, demonstrating their remarkable ability to maximize surface area and excel in chemical reactions. Imagine a material that is incredibly thin, almost like a sheet made up of individual atoms, where reactions can take place. The unique benefit arises from the extremely thin structure of 2DM, which gives them surface areas that are millions of times greater than their bulk counterparts [32]. Just like a physicist would do, let's consider the comparison between a tennis ball and a crumpled sheet of paper. Although they are made of the same material, the crumpled paper has a much larger surface area because of its folds and creases.

Active sites are of utmost importance within the vast surface area of 2DM. These active sites have distinct electronic and geometric configurations that make them ideal for hosting chemical reactions. With the abundance of exposed atoms in 2DM, there are numerous active sites available to facilitate reactions. This abundance results in multiple advantages, transforming 2DM into highly reactive substances. Understanding the multitude of active sites in 2DM offers numerous advantages, such as enhanced reaction rates, refined selectivity, and reduced activation energy [33]. By having a greater number of active sites at our disposal, we can process a larger quantity of reactant molecules at once, resulting in a significant boost to the overall reaction rate. Operating within the controlled environment of active sites enables precise adjustment of reaction pathways, facilitating the production of desired products while reducing the occurrence of undesired side reactions. In addition, the distinct electronic characteristics of 2DM play a role in lowering the energy needed to initiate reactions, thus boosting their catalytic performance. Although the increased surface area of 2DM is undeniably beneficial, it's important to acknowledge that it is not a cure-all solution. Factors like the intrinsic activity of the material and its stability under reaction conditions are also key players in determining their performance. Practical instances showcase the impressive catalytic abilities of 2DM in a range of applications [34]. As an example, nanosheets of Molybdenum disulfide (MoS_2) have the ability to efficiently convert water into hydrogen fuel using sunlight, because of their high surface area and active edge sites. In addition, researchers are investigating the potential of 2DM based catalysts to convert captured carbon dioxide into valuable chemicals or fuels, which could help mitigate climate change. With their impressive surface area and customizable properties, 2DM have emerged as exciting contenders for the future of batteries and fuel cells [35]. Their potential in electrocatalysis is truly remarkable. Research into 2DM and their surface area-driven applications is still in its early stages, but the possibilities for clean energy production, sustainable chemical processes, and advanced technologies are enormous. With a deep understanding of the intricate properties of these incredibly thin materials, we have the potential to shape a future that prioritizes clean energy and technological progress for the betterment of society as a whole.

2.4   Conductivity, stability, and charge separation efficiency in 2DM

Conductivity in 2DM varies greatly, with some demonstrating near-perfect metallic conductivity, like graphene, while others act as semiconductors or insulators. Understanding band structures is essential in this field, and physicists employ various techniques, such as doping, to adjust carrier concentration and improve conductivity. Furthermore, strain engineering utilizes mechanical strain to modify the lattice of the material, resulting in changes to the bandgap and conductivity [36]. By stacking different 2DM, the formation of heterostructures allows for the development of interfaces with customized electronic properties. This opens up exciting possibilities for the creation of new conductive pathways. Stability is a crucial factor to take into account when considering the practical applications of 2DM. Understanding the factors that affect stability involves considering the vulnerability to oxidation in the presence of air, which may require the use of protective coatings or carefully controlled environments. Stability can be influenced by the chemical composition, especially when it comes to certain 2DM that may degrade in specific chemical environments [37]. In addition, it is essential to preserve the atomically thin structure under different conditions to guarantee its structural integrity. Efficient charge separation is crucial for various applications, such as photovoltaics and photocatalysis. It depends on the material's capability to effectively separate electron-hole pairs generated by light. The importance of having a high surface area and customized band structures becomes clear when we consider the large number of active sites on the surface [38]. These sites provide ample opportunities for capturing light and creating electron-hole pairs. By harnessing tunable bandgaps, the charge separation efficiency is enhanced as built-in electric fields are generated, facilitating the separation of charges and effectively preventing recombination. These considerations highlight the complex nature of 2DM and their wide range of applications in various fields. Fig. 3 represents some of the 2D materials and their properties [39].

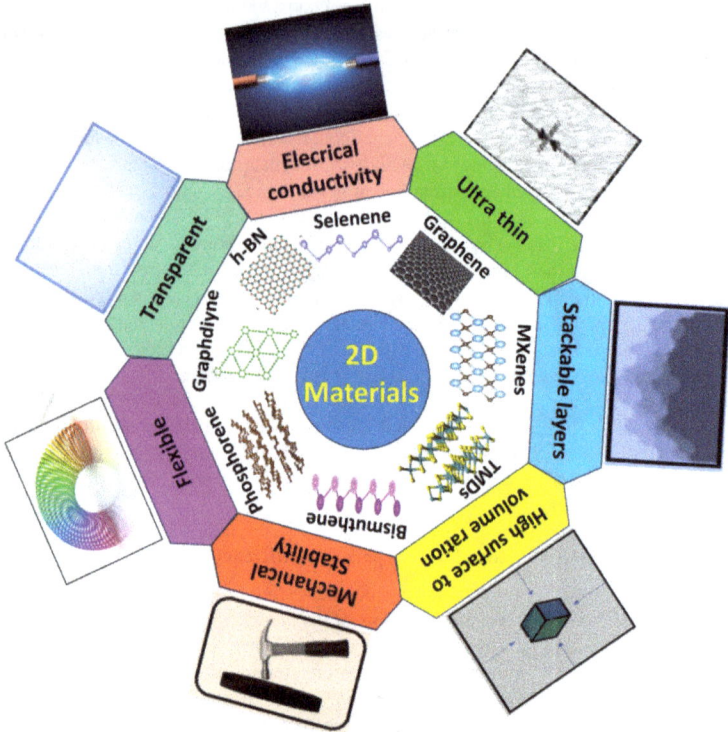

Fig. 3 2D materials and properties

3.   Applications of 2DM in Photocatalysis

3.1   Water splitting

Water splitting using sunlight has become an important research zone in the pursuit of clean and sustainable hydrogen fuel. Promising candidates in this field are 2DM photocatalysts, which possess unique properties. With its ultrathin structure, it offers a generous surface area that allows for optimal interaction with water molecules, leading to highly efficient reactions. By utilizing an extended light path within the 2D structure, the absorption of light is significantly enhanced, resulting in the optimization of solar energy capture. These materials possess tunable band gaps, which enable them to achieve optimal light absorption and charge separation [40]. Utilizing the lack of bulk material, direct charge transfer reduces energy losses and enhances photocatalytic efficiency. Ongoing research explores a range of 2DM photocatalysts, TMDs, graphene-based materials, MXenes, and Metal-Organic Frameworks (MOFs).

TMDs, such as MoS_2, WS_2, $MoSe_2$, and WSe_2, have been extensively researched as 2DM for water splitting. With their impressive band gaps, exceptional charge carrier mobility, and active edge sites, they enable a highly efficient process of hydrogen evolution [41]. Current research is centered on heterojunctions and defect engineering to develop catalytic activity and stability. Graphene-based materials are highly effective in supporting metal nanoparticles or other semiconductors in photocatalysis. Continuing research focuses on finding affordable techniques to apply metal nanoparticles to graphene and investigating the potential benefits of combining graphene with other 2DM, such as TMDs [42, 43]. MXenes, transition metal carbides, and nitrides are being studied for their exceptional electrical conductivity and adjustable band gaps. Researchers are exploring ways to enhance their water splitting capabilities, such as making surface modifications to improve stability in aqueous environments. Metal-Organic Frameworks (MOFs) are incredibly fascinating structures with remarkable porosity and the ability to be tailored for specific functions. These frameworks incorporate metal ions to create active sites that are ideal for facilitating hydrogen or oxygen evolution reactions [44]. Research is centered on the development of MOFs that have the perfect pore sizes and functionalities to effectively adsorb water and facilitate reactions. Additionally, there is a focus on combining these MOFs with other 2D materials such as TMDs to create hybrid photocatalysts.

Phosphides of various transition metals have gained significant attention as photocatalysts. They have demonstrated the capability to adsorb and release H_2 gas, making them highly effective in the hydrogen evolution reaction (HER), as supported by the research of Zhang et al. [45]. In addition, researchers have explored their ability to function as catalysts for both the HER and OER processes by incorporating metal hydroxo/oxo groups on the surfaces of the materials. This allows for the enhancement of both processes through a co-catalysis mechanism. Researchers have extensively studied Ni, Co and Fe phosphides, for their potential in photocatalytic applications. The results have been highly promising. Understanding the phosphides content of the metal catalyst material is crucial in determining the HER activity. It plays a significant role in the catalyst's interaction and affinity with H atoms. This has been observed by Pan et al. [46] and Anantharaj et al., [47], who have established the correlation between the catalytic activity of HER and the phosphides. In other words, increasing the Phosphides ratio leads to a proportional increase in both activity and release rate. A study conducted by Ledendecker, et al., [48] explored the potential of nickel phosphide as a dual functional photocatalyst. At 1.7 V, the measured activities were noteworthy for both OER and HER, attaining a current-density of 10 mA cm^{-2}. The nanostructure of Ni_5P_4 was created through a straightforward method of heating elemental phosphorus and a Nickel foil sheet in an inert environment. The material displayed remarkable performance in the particularly important OER kinetics, a task that is often challenging in the field of photocatalysis. The development of a very active NiOOH deposition on the surface could be the cause of this characteristic. During a 20-hour test in an acidic, this deposition aids in lowering the operation over-potential, shifting the electronic structure, increasing the current density, and maintaining stable operation. Stern et al. offered Ni_2P as a possible electrode catalyst [49]. This active species was found to enhance both the HER and OER kinetics, making it a cost-effective and efficient alternative catalyst made from a single material. The Ni_2P nanoparticles and nanowires were created by heating the precursor ingredients, resulting in similar OER activities. Through detailed structural studies, it was revealed that the $NiOx/Ni_2P$ nanoparticles possess a shell/core structure this distinctive arrangement plays a crucial role in the process, as the core of Ni2P serves as an electronic pathway to the shell of NiOx. Consequently, these nanoparticles demonstrate superior diffusive properties and enhanced activity when compared to NiOx nanoparticles of similar size. According to the reported values of current density (10 mA/cm^2) achieved with an overpotential as low as 290 mV for OER and 10 mA/cm2 with a 136 mV overpotential for HER in acidic medium, the produced catalyst material is moving toward widespread water splitting photocatalytic systems as a viable and workable candidate material for reactions. In addition to nickel, researchers have also explored other transition metal phosphides for their catalytic properties in releasing hydrogen and oxygen from water when exposed to light. Researchers and reviewers have extensively studied the potential of various forms and combinations of cobalt phosphides, which have shown promising prospects.

MOFs consist of metal centers linked to organic bridging ligands, leading to a variety of compositions, geometries, and pore spacing depending on the synthesis procedure, conditions, and initiators. The unique metal-centric interlinked geometric structure enables MOFs to be used in a multitude of applications across different industries. These include drug delivery systems, gas adsorption, optics, separation, catalysis, and sensors [50]. Researchers are constantly examining the possibilities and opportunities for their efficient utilization and improvements. Recently, there has been a surge in the development of catalyst materials based on MOFs for photocatalytic water splitting reactions. The goal is to create new and practical bi-functional promoters for solar hydrogen production from water. Incorporating light-sensitive building blocks, such as aminoterephthalates and porphyrins, into the MOF backbone is a common practice. Another approach is to enhance catalytic capacities by doping with active metals or noble metals. Additionally, inorganic nanoparticles can be encapsulated in MOFs, and ternary composites can be assembled. Introducing electron donors or trapping materials can help address the issue of charge recombination. In this context, Lan et al. have made a significant contribution by integrating the widely recognized high reactivity of noble metals in water splitting systems into materials based on MOFs. They have successfully generated three types of Pt-tempered heterojunctions: Pt-Zn_3P_2-CoP, Pt-ZnS-CoS, and Pt-ZnO-Co_3O_4 [51].

3.2 CO_2 reduction

In recent times, there has been a lot of interest in 2D metal chalcogenide nanosheets such as MoS_2, WS_2, SnS_2, TiS_2, $MoSe_2$, WSe_2, and others. These nanosheets have caught the attention of researchers because of their remarkable catalytic activity and cost-effectiveness when compared to noble metals. One particularly interesting nanosheet is MoS_2, which consists of hexagonal layers of Mo atoms sandwiched between two layers of sulfur atoms. This unique composition gives MoS_2 its potential as a semiconductor, with a band gap that varies depending on its thickness. Additionally, MoS_2 is abundant in nature, specifically in the form of molybdenite [52]. Although bulk MoS_2 is a semiconductor with an indirect band gap of approximately 1.3 eV, which restricts its application in photocatalytic reactions due to its insufficient oxidation/reduction potential, isolated MoS_2 nanosheets possess a direct bandgap of 1.96 eV. This characteristic grants them exceptional photoinduced catalytic capability. MoS_2 is often used as a catalyst or non-noble metal co-catalyst along with other photocatalysts to create heterojunctions, which boost photocatalytic activity for applications such as photodegradation and water splitting. In the realm of transition metal dichalcogenide (TMD) nanosheets, certain types like WS_2 and NiS_2 have been found to be economical catalysts or co-catalysts. However, there is limited research on their ability to reduce CO_2 through photocatalysis. Recently, Tu et al. delved into this area and successfully created 2D MoS_2-TiO_2 hybrid nanojunctions. These nanojunctions exhibited impressive photocatalytic activity for converting CO_2 in aqueous solution [53]. In a study conducted by Dai et al., they discovered that the combination of MoS_2 nanosheets and hierarchical flower-like Bi_2WO_6 nanocomposites resulted in a remarkable improvement in photocatalytic CO_2 reduction under visible light [54]. This finding highlights the potential of MoS_2 as a co-catalyst, as it enhances light absorption and charge transfer, ultimately boosting the photocatalytic activity of the material. The proposed photocatalytic mechanism explores reactive substrates found in CO_2 aqueous solution, indicating the thermodynamic feasibility of CO_2 photoreduction.

During the past decades, there has been a lot of interest in bismuth oxyhalides, specifically BiOX (X = Cl, Br, and I), due to their remarkable layered structure and impressive optical and electrical properties. With its layered structure, bismuth oxyhalides exhibit remarkable photocatalytic activity. This is due to the presence of self-built internal static electric fields, which are formed by the arrangement of $[Bi_2O_2]^{2+}$ layers between halogen ion slabs. Specifically, 2D nanosheets of BiOX (X = Cl, Br, and I) and 3D hierarchical frameworks containing 2D nanoflakes demonstrated exceptionally high photocatalytic activity for the elimination of pollutants from air, water, and biological contaminants. As an example, Wang et al. created porous BiOCl micro-flowers made up of ultrathin nanosheets (4 nm) that had a high percentage of {001} facets exposed. These micro-flowers exhibited a powerful photooxidative capability to efficiently degrade dye pollutants (RhB, MB, MO) without any complications [55]. In a groundbreaking discovery, Guan and their colleague observed remarkable solar photocatalytic activity in ultrathin BiOCl nanosheets (2.7 nm) that showcased fully exposed active {001} facets. This breakthrough led to significant advancements in the field of photodegradation of RhB [56]. Unfortunately, the photocatalytic reduction activity was relatively poor due to the high positive conduction band minimum (CBM) of BiOX photocatalysts, and there were very few studies on the photoreduction of CO_2 or H_2O employing BiOX.

Photocatalysts made from metal oxides are frequently used to help reduce carbon dioxide (CO_2). Several catalysts include transition metal cations like Ti^{4+}, Zr^{4+}, Nb^{5+}, Ta^{5+}, W^{6+}, and Mo^{6+}, which have a d_0 configuration. TiO_2 is widely researched because of its remarkable stability, lack of toxicity, affordability, and widespread availability. It is worth mentioning that 2D nanosheets made from TiO_2, which are obtained by exfoliating layered titanate, have attracted a great deal of attention in research [57]. In a remarkable study, Xu et al. achieved the synthesis of anatase TiO_2 single crystals consisting of ultrathin TiO_2 nanosheets measuring 2 nm in thickness, which revealed an impressive 95% exposure of (100) facets. These nanosheets showed a significant fivefold boost in activity for both the hydrogen evolution reaction (HER) and CO_2 reduction when compared to TiO_2 cuboids with 53% of exposed (100) facets [58]. This highlights the potential of TiO2 nanosheets to improve performance in CO_2 reduction reactions, highlighting their efficiency and unique structural characteristics.

Metal sulfides, in contrast to their oxide counterparts, possess valence bands that exhibit a greater concentration of S3p character and narrower band gaps. However, there is widespread worry regarding the restricted energy of the photo-generated holes in their valence bands. This limited energy may not be enough to effectively oxidize water and could potentially lead to irreversible photocorrosion. To address this issue and improve their stability, hole scavengers are frequently introduced. One well-known photocatalyst for visible light is CdS, which has a band gap of 2.4 eV that perfectly matches the solar spectrum [59]. The first investigation on the photocatalytic performance of CdS in the carbon dioxide (CO_2) reduction reaction under visible light was published in 1988 by Eggins et al. They witnessed the creation of different primary products, such as formaldehyde, methanol, formate, acetate, and glyoxylate [60]. In a separate study, Tu et al. performed an experiment where they employed a method to cultivate 2D MoS_2 nanosheets directly on TiO_2 nanosheets, leading to the creation of 2D hybrid nanojunctions. Through the strategic arrangement of MoS_2 nanosheets in direct contact with TiO_2, the aim was to maximize the interfacial area and optimize the catalytic efficiency and overall performance of the photocatalyst. [61].

3.3   Pollutant degradation

MoS_2 materials have been widely acknowledged for their excellent performance in degrading contaminants through photocatalysis. They have strong light absorption, superior photocatalytic capabilities, and are cost-effective [62]. MoS_2 thin films with perpendicularly aligned layers are quite remarkable. These films have a multitude of 2D sites thanks to their well-developed dangling bonds. As a result, they display exceptional chemical reactivity, making them highly effective in breaking down harmful compounds [63]. In recent research, scientists have been working on creating TiO_2/MoS_2 nanocomposites to prevent charge recombination and improve absorption in the visible region [64]. In addition, researchers have used the deposition of CDs and Ag_3PO_4 NPs onto MoS_2 sheets to enhance their ability to degrade photosensitive materials. This method effectively separates the electron-hole pairs produced during the photodegradation process [65]. MOFs have been extensively used in a wide range of environmental applications, particularly in the efficient degradation of organic dyes through photodegradation [66]. The initial publication regarding a MOF's ability to adsorb methyl orange (MO) was released in 2010, while its documented success in degrading phenol through photocatalysis dates back to 2007 [67]. The photodegradation process offers a distinct advantage over adsorption due to its heightened efficiency and capability to fully decompose the intended product. Composites that utilize MOFs, especially those incorporating polyoxometalate (POM) structures, have demonstrated promising capabilities in the degradation of organic dyes [68]. It's worth noting that catalysts containing Keggin-type clusters have shown promise in POM-based systems. Nevertheless, a comprehensive examination of their design and the correlations between their structure and activity is still required [69]. Using basic POM-based photocatalysts can be quite challenging because of their relatively low stability and limited ability to absorb light.

Jiang et al. studied the MIL-53(Fe)/visible light/H_2O_2 system to investigate the synergistic effect. They analyzed the band position, which is directly related to the redox capacity of photoinduced charge carriers. The flatband potential of MIL-53(Fe) was determined using a conventional Mott-Schottky plot at a frequency of 100 Hz in the absence of light. It was found to be approximately 0.36 V vs. NHE, which is more negative than the redox potential of $O_2/\bullet O_2^-$ (0.33 V vs. NHE). The conduction band (CB) of MIL-53(Fe) was predicted to have a potential of 0.46 V vs. NHE [70, 71]. The redox potential of RhB was determined to be around 1.43 V vs. NHE, which is lower than the VB level of MIL-53(Fe). This suggests that the direct hole oxidation process is energetically favorable. By employing the photoluminescence (PL) technique, the presence of ·OH radicals in the MIL-53(Fe)/visible light/H_2O_2 catalytic system was verified. When comparing the results of MIL-53(Fe)/visible light and MIL-53(Fe)/H_2O_2 systems, it was found that the MIL-53(Fe)/visible light/H_2O_2 system had a stronger PL intensity. This indicates that more ·OH radicals were generated, which can be attributed to the synergistic effects of combining MIL-53(Fe) and H_2O_2.

In a study by Guesh et al., [72] MIL-100(Fe) was synthesized to investigate its effectiveness in breaking down methyl orange (MO) dye in water. The photocatalytic process involved exposing the sample to both ultraviolet and solar light radiation. Unlike conventional hydrothermal methods, this synthesis took place at room temperature in just a few hours, without the use of harsh inorganic acids, and eliminated the requirement for high-temperature washing after synthesis. After 7 hours of exposure to UV light, it was found that only 64% of the 5 ppm MO underwent photo-catalytic degradation. Interestingly, solar light managed to achieve a 40% degradation for the same concentration of MO. As part of another study, Mahmoodi et al. [73] conducted an experiment involving three MIL-100(Fe) samples: MIL-100-1, MIL-100-2, and MIL-100-3. They used a UV-C lamp (Philips 9 W) as the source of irradiation to degrade Basic Blue 41. The reaction exhibited a first-order kinetics model, and the catalyst showed the ability to be recovered and reused for three cycles. During three cycles, MIL-100-1 demonstrated exceptional photocatalytic activity, with no noticeable decline in performance. In their study, Mahmoodi et al. [74] investigated the degradation of Basic Blue 41 using a Cu-based metal-organic framework (MOF). MOF-199 was synthesized using solvothermal conditions at 140 °C for 8 hours. The degradation efficiency achieved through UV lamp (9 W) irradiation was remarkably high, reaching approximately 99%. In addition, Liu et al. [75] synthesized MIL-53(Fe) using the solvothermal method with different concentrations and temperatures. The main objective was to study the process of photocatalytic degradation of RhB dye using visible light from a 500 W Xe lamp. The results showed that the MOF created with the ideal combination of $FeCl_3·6H_2O$ and H2BDC at a ratio of 1:2 and a temperature of 150 °C demonstrated the most effective degradation process.

Conclusion

The intriguing properties and potential applications of 2DM have recently generated considerable attention in the field of photocatalysis. With their novel properties and surprising behaviours, they challenge long-held scientific assumptions and provide fresh insights into the world. When it comes to light-driven chemical processes, 2DM really shine because of their exceptional advantages. Studying the bonding and coordination of 2DM is different from bulk metals due to the weaker interlayer interactions, such as van der Waals forces or covalent bonds. This characteristic enables easy exfoliation and stacking into heterostructures. The lower coordination number in 2DM, where each atom connects with fewer neighbours, has a significant impact on attributes such as conductivity and magnetism. When electrons are confined within a single plane, their optical properties, electrical band structures, and energies undergo significant changes. Understanding the behaviour of 2DM requires a thorough consideration of the significant influence of surface effects. In addition, the distinct electrical characteristics of 2DM improve their catalytic efficiency by reducing the energy needed to start reactions. While the increased surface area of 2D metals does offer certain advantages, it's crucial to acknowledge that it doesn't provide a solution for every issue. Understanding the activity and stability of materials under reaction conditions is essential. The composition, geometry, and pore spacing of metal-organic frameworks (MOFs) can vary greatly due to the diverse range of synthesis procedures, environmental factors, and initiators utilized. MOFs possess unique properties due to their interconnected geometries centred on metal

Emerging Materials for Next Frontier Energy and Environment Applications   Materials Research Forum LLC
Materials Research Foundations 170 (2024) 21-40   https://doi.org/10.21741/9781644903292-2

atoms. This makes them highly versatile and applicable in various fields including optics, gas adsorption, separation, catalysis, and sensors. Scientists are constantly exploring the possibilities for practical applications and improvements of these materials. Molybdenite, a type of MoS_2, is a naturally occurring mineral that is widely available. Although bulk MoS_2 has a band gap of approximately 1.3 eV, which restricts its effectiveness in photocatalytic processes due to insufficient oxidation/reduction potential, the situation is different for isolated MoS_2 nanosheets. These nanosheets possess a direct band gap of 1.96 eV, leading to an impressive photoinduced catalytic capacity. When MoS_2 is used as a catalyst or non-noble metal co-catalyst with other photocatalysts, heterojunctions are formed to boost photocatalytic activity for applications such as photodegradation and water splitting. WS_2 and NiS_2 are considered to be highly cost-effective among the different transition metal dichalcogenide (TMD) nanosheets. Photodegradation has a clear advantage over adsorption because of its superior efficiency and ability to completely break down the target product. Composites utilizing MOFs, especially those with polyoxometalate (POM) structures, have demonstrated promise in the degradation of organic dyes. The unique characteristics of two-dimensional materials such as large surface area, variable electrical properties, and high charge carrier mobility make them important for the development of photocatalytic technologies. These characteristics open the door to significantly better environmental and energy-related applications by enabling more effective, reliable, and adaptable photocatalytic devices.

Reference

[1] Y. Shi and B. Zhang, Recent advances in transition metal phosphide nanomaterials: synthesis and applications in hydrogen evolution reaction, Chem. Soc. Rev. 45 6 (2016) 1529-1541. https://doi.org/10.1039/C5CS00434A

[2] M. A. Hassaan, M. A. El-Nemr, M. R. Elkatory, S. Ragab, V. C. Niculescu, and A. El Nemr, Principles of photocatalysts and their different applications: a review, Top. Curr. Chem. 381 6 (2023) 31. https://doi.org/10.1007/s41061-023-00444-7

[3] Z. Kuspanov, B. Baglan, A. Baimenov, A. Issadykov, M. Yeleuov, and C. Daulbayev, Photocatalysts for a sustainable future: Innovations in large-scale environmental and energy applications, Sci. Total Environ. (2023) 163914. https://doi.org/10.1016/j.scitotenv.2023.163914

[4] R. Roshan Chandrapal, S. Bharathkumar, G. Bakiyaraj, V. Ganesh, J. Archana, and M. Navaneethan. Hydrothermally synthesized strontium-modified ZnO hierarchical nanostructured photocatalyst for second-generation fluoroquinolone degradation. Appl. Nanosci. 12 6 (2022) 1869-1884. https://doi.org/10.1007/s13204-022-02414-9

[5] J. Zhang, H. Wu, L. Shi, Z. Wu, S. Zhang, S. Wang, and H. Sun, Photocatalysis coupling with membrane technology for sustainable and continuous purification of wastewater, Sep. Purif. Tech. 329 (2023) 125225. https://doi.org/10.1016/j.seppur.2023.125225

[6] K. S. Novoselov, A. K. Geim, S. V. Morozov, D. Jiang, Y. Zhang, S. V. Dubonos, I. V. Grigorieva, and A. A. Firsov, Electric field effect in atomically thin carbon films, sci. 306 5696 (2004) 666-669. https://doi.org/10.1126/science.1102896

[7] W. Choi, N. Choudhary, G. H. Han, J. Park, D. Akinwande, and Y. H. Lee, Recent development of two-dimensional transition metal dichalcogenides and their applications, Mater. Today 20 3 (2017) 116-130. https://doi.org/10.1016/j.mattod.2016.10.002

[8] H. Zhou, J. R. Long, and O. M. Yaghi, Introduction to metal-organic frameworks, Chem. Rev. 112 2 (2012) 673-674. https://doi.org/10.1021/cr300014x

[9] Y. Gogotsi, and B. Anasori, The rise of MXenes, ACS nano 13 8 (2019) 8491-8494. https://doi.org/10.1021/acsnano.9b06394

[10] Y. Li, C. Gao, R. Long, and Y. Xiong, Photocatalyst design based on two-dimensional materials, Mater. Today chem. 11 (2019) 197-216. https://doi.org/10.1016/j.mtchem.2018.11.002

[11] S. K. Chakraborty, B. Kundu, B. Nayak, S. P. Dash, and P. K. Sahoo, Challenges and opportunities in 2D heterostructures for electronic and optoelectronic devices, iScience 25 3 (2022) 103942. https://doi.org/10.1016/j.isci.2022.103942

[12] D. S. Schulman, A. J. Arnold, and S. Das, Contact engineering for 2D materials and devices, Chem. Soc. Rev. 47 9 (2018) 3037-3058. https://doi.org/10.1039/C7CS00828G

[13] K. Jiang, and Q. Weng, Miniaturized Energy Storage Devices Based on Two-Dimensional Materials, ChemSusChem 13 6 (2020) 1420-1446. https://doi.org/10.1002/cssc.201902520

[14] Q. R. H. Deng, P. Li, H. Gomaa, S. Wu, C. An, and N. Hu, Research Progress of transition-metal dichalcogenides for hydrogen evolution reaction, J. Mater. Chem. A 11 (2023) 24434-24453. https://doi.org/10.1039/D3TA04475K

[15] N. Kumar, R. Salehiyan, V. Chauke, O. J. Botlhoko, K. Setshedi, M. Scriba, M. Masukume, and S. S. Ray, Top-down synthesis of graphene: A comprehensive review, FlatChem 27 (2021) 100224. https://doi.org/10.1016/j.flatc.2021.100224

[16] M. Jakhar, A. Kumar, P. K. Ahluwalia, K. Tankeshwar, and R. Pandey, Engineering 2D materials for photocatalytic water-splitting from a theoretical perspective, Mater. 15 6 (2022) 2221. https://doi.org/10.3390/ma15062221

[17] G. Liao, C. Li, S. Liu, B. Fang, and H. Yang, Z-scheme systems: From fundamental principles to characterization, synthesis, and photocatalytic fuel-conversion applications, Phys. Rep. 983 (2022) 1-41. https://doi.org/10.1016/j.physrep.2022.07.003

[18] J. Lipton, G. Weng, J. A. Röhr, H. Wang, and A. D. Taylor, Layer-by-layer assembly of two-dimensional materials: meticulous control on the nanoscale, Matter 2 5 (2020) 1148-1165. https://doi.org/10.1016/j.matt.2020.03.012

[19] M. I. Nabeel, D. Hussain, N. Ahmad, M. Najam-ul-Haq, and S. G. Musharraf, Recent Advancements in Fabrication and Photocatalytic Applications of Graphitic Carbon Nitride-Tungsten Oxide Nanocomposites, Nanoscale Adv. 5 (2023) 5214-5255. https://doi.org/10.1039/D3NA00159H

[20] X. Li, J. Zhu, and B. Wei, Hybrid nanostructures of metal/two-dimensional nanomaterials for plasmon-enhanced applications, Chem. Soc. Rev. 45 11 (2016) 3145-3187. https://doi.org/10.1039/C6CS00195E

[21] T. Wang, M. Park, Q. Yu, J. Zhang, and Y. Yang, Stability and synthesis of 2D metals and alloys: a review, Mater. Today Adv. 8 (2020) 100092. https://doi.org/10.1016/j.mtadv.2020.100092

[22] K. Bharathi, K. Sathiyamoorthy, G. Bakiyaraj, J. Archana, and M. Navaneethan. 2D V2O5 nanoflakes on 2D p-gC3N4 nanosheets of heterostructure photocatalysts with the enhanced photocatalytic activity of organic pollutants under direct sunlight, Surf. Interf. 41 (2023) 103219. https://doi.org/10.1016/j.surfin.2023.103219

[23] A. B. Kaul, Two-dimensional layered materials: Structure, properties, and prospects for device applications, J. Mater. Res. 29 3 (2014) 348-361. https://doi.org/10.1557/jmr.2014.6

[24] P. Hess, Bonding, structure, and mechanical stability of 2D materials: the predictive power of the periodic table, Nanoscale Horiz. 6, 11 (2021) 856-892. https://doi.org/10.1039/D1NH00113B

[25] V. Shanmugam, R. A. Mensah, K. Babu, S. Gawusu, A. Chanda, Y. Tu, R. E. Neisiany, M. Försth, G. Sas, and O. Das, A review of the synthesis, properties, and applications of 2D materials, Part. Part. Syst. Charact. 39, no. 6 (2022) 2200031. https://doi.org/10.1002/ppsc.202200031

[26] S. Ahmed, and J. Yi, Two-dimensional transition metal dichalcogenides and their charge carrier mobilities in field-effect transistors, Nano-Micro let. 9 (2017) 1-23. https://doi.org/10.1007/s40820-016-0103-7

[27] R. K. Mishra, K. Verma, I. Chianella, H. Y. Nezhad, and S. Goel, Borophene: A 2D Wonder Shaping the Future of Nanotechnology and Materials Science, http://dx.doi.org/10.13140/RG.2.2.30733.51680

[28] Y. Wang, B. Ren, J. Z. Ou, K. Xu, C. Yang, Y. Li, and H. Zhang, Engineering two-dimensional metal oxides and chalcogenides for enhanced electro-and photocatalysis, Sci. Bull. 66 12 (2021) 1228-1252. https://doi.org/10.1016/j.scib.2021.02.007

[29] H. Liu, B. Xu, J-M. Liu, J. Yin, F. Miao, C. Duan, and X. G. Wan, Highly efficient and ultrastable visible-light photocatalytic water splitting over ReS 2, Phys. Chem. Chem. Phys. 18 21 (2016) 14222-14227. https://doi.org/10.1039/C6CP01007E

[30] N. Mphuthi, L. Sikhwivhilu, and S. S. Ray, Functionalization of 2D MoS2 nanosheets with various metal and metal oxide nanostructures: their properties and application in electrochemical sensors, Biosensors 12 6 (2022) 386. https://doi.org/10.3390/bios12060386

[31] Q. Bi, K. Hu, J. Chen, Y. Zhang, M. S. Riaz, J. Xu, Y. Han, and F. Huang, Black phosphorus coupled black titania nanocomposites with enhanced sunlight absorption properties for efficient photocatalytic CO2 reduction, Appl. Catal. B: Environ. 295 (2021) 120211. https://doi.org/10.1016/j.apcatb.2021.120211

[32] B. Zhang, T. Fan, N. Xie, G. Nie, and H. Zhang, Versatile Applications of Metal Single-Atom@ 2D Material Nanoplatforms, Adv. Sci. 6 21 (2019) 1901787. https://doi.org/10.1002/advs.201901787

[33] H. Liu, M. Cheng, Y. Liu, J. Wang, G. Zhang, L. Li, L. Du, G. Wang, S. Yang, and X. Wang, Single atoms meet metal-organic frameworks: collaborative efforts for efficient

photocatalysis, Energy Environ. Sci. 15 9 (2022) 3722-3749.
https://doi.org/10.1039/D2EE01037B

[34] S. S. Varghese, S. H. Varghese, S. Swaminathan, K. K. Singh, and V. Mittal, Two-dimensional materials for sensing: graphene and beyond, Electronics 4 3 (2015) 651-687.
https://doi.org/10.3390/electronics4030651

[35] N. S. Powar, C. B. Hiragond, D. Bae, and S. In, Two-dimensional metal carbides for electro-and photocatalytic CO2 reduction, J. CO2 Util. 55 (2022) 101814.
https://doi.org/10.1016/j.jcou.2021.101814

[36] Z. Peng, X. Chen, Y. Fan, D. J. Srolovitz, and D. Lei, Strain engineering of 2D semiconductors and graphene: from strain fields to band-structure tuning and photonic applications, Light Sci. Appl. 9 (2020) 190. https://doi.org/10.1038/s41377-020-00421-5

[37] P. Hess, Bonding, structure, and mechanical stability of 2D materials: the predictive power of the periodic table, Nanoscale Horiz.s 6 11 (2021) 856-892.
https://doi.org/10.1039/D1NH00113B

[38] H. Yang, B. Yang, W. Chen, and J. Yang, Preparation and photocatalytic activities of TiO2-based composite catalysts, Catalysts 12 10 (2022) 1263.
https://doi.org/10.3390/catal12101263

[39] M. Srivastava, S. Banerjee, S. Bairagi, P. Singh, B. Kumar, P. Singh, R. D. Kale, D. M. Mulvihill, and S. Wazed Ali, Recent progress in molybdenum disulfide (MoS2) based flexible nanogenerators: An inclusive review, Chem. Eng. J. (2023) 147963.
https://doi.org/10.1016/j.cej.2023.147963

[40] B. Fang, Z. Xing, D. Sun, Z. Li, and W. Zhou, Hollow semiconductor photocatalysts for solar energy conversion, Adv. Powder Mater. 2 (2022) 100021.
https://doi.org/10.1016/j.apmate.2021.11.008

[41] S. Li, Z. Zhao, D. Yu, J. Zhao, Y. Su, Y. Liu, Y. Lin, W. Liu, H. Xu, and Z. Zhang, Few-layer transition metal dichalcogenides (MoS2, WS2, and WSe2) for water splitting and degradation of organic pollutants: Understanding the piezocatalytic effect, Nano Energy 66 (2019) 104083. https://doi.org/10.1016/j.nanoen.2019.104083

[42] J. Raveena, V. S. Manikandan, G. Bakiyaraj, and M. Navaneethan. Co substituted SnS 2 nanoflakes performed as cost-effective counter electrode for DSSCs applications. Journal of Materials Science: Materials in Electronics (2022) 1-8. https://doi.org/10.1007/s10854-021-07015-w

[43] R. Roshan Chandrapal, G. Bakiyaraj, S. Bharathkumar, V. Ganesh, J. Archana, M. Navaneethan, Novel combustion technique synthesis of ZnCo2O4 nanobeads/g-C3N4 nanosheet heterojunction photocatalyst in-effect of enhancing photocatalytic degradation of environmental remediation, Surf. Interf. 41 (2023) 103269.
https://doi.org/10.1016/j.surfin.2023.103269

[44] V. F. Yusuf, N. I. Malek, and S. K. Kailasa, Review on Metal-Organic Framework Classification, Synthetic Approaches, and Influencing Factors: Applications in Energy, Drug Delivery, and Wastewater Treatment, ACS omega 7 49 (2022) 44507-44531.
https://doi.org/10.1021/acsomega.2c05310

[45] L. Zhang, Z. Shi, Y. Lin, F. Chong, and Y. Qi, Design strategies for large current density hydrogen evolution reaction, Front. Chem. 10 (2022) 866415. https://doi.org/10.3389/fchem.2022.866415

[46] Y. Pan, Y. Liu, J. Zhao, K. Yang, J. Liang, D. Liu, W. Hu, D. Liu, Y. Liu, and C. Liu, Monodispersed nickel phosphide nanocrystals with different phases: synthesis, characterization and electrocatalytic properties for hydrogen evolution, J. Mater. Chem. A 3 (2015) 1656-1665. https://doi.org/10.1039/C4TA04867A

[47] S. Anantharaj, S. R. Ede, K. Sakthikumar, K. Karthick, S. Mishra, and S. Kundu, Recent trends and perspectives in electrochemical water splitting with an emphasis on sulfide, selenide, and phosphide catalysts of Fe, Co, and Ni: a review, ACS Catalysis 6 12 (2016) 8069-8097. https://doi.org/10.1021/acscatal.6b02479

[48] M. Ledendecker, S. K. Calderón, C. Papp, H. Steinrück, M. Antonietti, and M. Shalom, The synthesis of nanostructured Ni5P4 films and their use as a non-noble bifunctional electrocatalyst for full water splitting, Angewandte Chemie International Edition 54, no. 42 (2015) 12361-12365. https://doi.org/10.1002/anie.201502438

[49] L. Stern, L. Feng, F. Song, and X. Hu, Ni 2 P as a Janus catalyst for water splitting: the oxygen evolution activity of Ni 2 P nanoparticles, Energy Environ. Sci. 8 (2015) 2347-2351. https://doi.org/10.1039/C5EE01155H

[50] S. Dutt, A. Kumar, and S. Singh, Synthesis of Metal Organic Frameworks (MOFs) and Their Derived Materials for Energy Storage Applications, Clean Technol. 5 (2023) 140-166. https://doi.org/10.3390/cleantechnol5010009

[51] M. Lan, R. Guo, Y. Dou, J. Zhou, A. Zhou, and J. Li, Fabrication of porous Pt-doping heterojunctions by using bimetallic MOF template for photocatalytic hydrogen generation, Nano Energy 33 (2017) 238-246. https://doi.org/10.1016/j.nanoen.2017.01.046

[52] B. Balan, M. M. Xavier, and S. Mathew, MoS2-Based Nanocomposites for Photocatalytic Hydrogen Evolution and Carbon Dioxide Reduction, ACS omega 8 (2023) 25649-25673. https://doi.org/10.1021/acsomega.3c02084

[53] W. Tu, Y. Li, L. Kuai, Y. Zhou, Q. Xu, H. Li, X. Wang, M. Xiao, and Z. Zou, Construction of unique two-dimensional MoS2-TiO2 hybrid nanojunctions: MoS2 as a promising cost-effective cocatalyst toward improved photocatalytic reduction of CO2 to methanol, Nanoscale 9, no. 26 (2017) 9065-9070. https://doi.org/10.1039/C7NR03238B

[54] W. Dai, J. Yu, Y. Deng, X. Hu, T. Wang, and X. Luo, Facile synthesis of MoS2/Bi2WO6 nanocomposites for enhanced CO2 photoreduction activity under visible light irradiation, Appl. Surf. Sci. 403 (2017) 230-239. https://doi.org/10.1016/j.apsusc.2017.01.171

[55] D. Wang, G. Gao, Y. Zhang, L. Zhou, A. Xu, and W. Chen, Nanosheet-constructed porous BiOCl with dominant {001} facets for superior photosensitized degradation, Nanoscale 4, no. 24 (2012) 7780-7785. https://doi.org/10.1039/c2nr32533k

[56] M. Guan, C. Xiao, J. Zhang, S. Fan, R. An, Q. Cheng, J. Xie, M. Zhou, B. Ye, and Y. Xie, Vacancy associates promoting solar-driven photocatalytic activity of ultrathin bismuth oxychloride nanosheets, J. American Chem. Soc. 135, no. 28 (2013) 10411-10417. https://doi.org/10.1021/ja402956f

[57] Z. U. Rehman, M. Bilal, J. Hou, F. K. Butt, J. Ahmad, S. Ali, and A. Hussain, Photocatalytic CO2 reduction using TiO2-based photocatalysts and TiO2 z-scheme heterojunction composites: A review, Molecules 27, no. 7 (2022) 2069. https://doi.org/10.3390/molecules27072069

[58] H. Xu, S. Ouyang, P. Li, T. Kako, and J. Ye, Response to Comment on High-Active Anatase TiO2 Nanosheets Exposed with 95%{100} Facets Toward Efficient H2 Evolution and CO2 Photoreduction, ACS Appl. Mater. Interf. 5 17 (2013) 8262-8262. https://doi.org/10.1021/am402298g

[59] X. Chen, S. Shen, L. Guo, and S. S. Mao, Semiconductor-based photocatalytic hydrogen generation, Chem. Rev. 110 11 (2010) 6503-6570. https://doi.org/10.1021/cr1001645

[60] B. R. Eggins, P. KJ Robertson, E. P. Murphy, E. Woods, and J. TS Irvine, Factors affecting the photoelectrochemical fixation of carbon dioxide with semiconductor colloids, J. Photochem. Photobio. A: Chem. 118 1 (1998) 31-40. https://doi.org/10.1016/S1010-6030(98)00356-6

[61] W. Tu, Y. Li, L. Kuai, Y. Zhou, Q. Xu, H. Li, X. Wang, M. Xiao, and Z. Zou, Construction of unique two-dimensional MoS2-TiO2 hybrid nanojunctions: MoS2 as a promising cost-effective cocatalyst toward improved photocatalytic reduction of CO2 to methanol, Nanoscale 9 26 (2017) 9065-9070. https://doi.org/10.1039/C7NR03238B

[62] M. Wu, L. Li, N. Liu, D. Wang, Y. Xue, and L. Tang, Molybdenum disulfide (MoS2) as a co-catalyst for photocatalytic degradation of organic contaminants: A review, Process Saf. Environ. Prot. 118 (2018) 40-58. https://doi.org/10.1016/j.psep.2018.06.025

[63] M. A. Islam, J. Church, C. Han, H. Chung, E. Ji, J. H. Kim, N. Choudhary, G. Lee, W. Hyoung Lee, and Y. Jung, Noble metal-coated MoS2 nanofilms with vertically-aligned 2D layers for visible light-driven photocatalytic degradation of emerging water contaminant, Sci. Rep. 7 1 (2017) 14944. https://doi.org/10.1038/s41598-017-14816-9

[64] E. Rahmanian, R. Malekfar, and M. Pumera, Nanohybrids of Two-Dimensional Transition-Metal Dichalcogenides and Titanium Dioxide for Photocatalytic Applications, Chem. A Eur. J. 24 1 (2018) 18-31. https://doi.org/10.1002/chem.201703434

[65] N. Li, Z. Liu, M. Liu, C. Xue, Q. Chang, H. Wang, Y. Li, Z. Song, and S. Hu, Facile synthesis of carbon dots@ 2D MoS2 heterostructure with enhanced photocatalytic properties, Inorg. Chem. 58 9 (2019) 5746-5752. https://doi.org/10.1021/acs.inorgchem.9b00111

[66] Q. Sun, L. Qin, C. Lai, S. Liu, W. Chen, F. Xu, D. Ma, Constructing functional metal-organic frameworks by ligand design for environmental applications, J. Hazard. Mater. 447 (2023) 130848. https://doi.org/10.1016/j.jhazmat.2023.130848

[67] E. Haque, J. W. Jun, and S. H. Jhung, Adsorptive removal of methyl orange and methylene blue from aqueous solution with a metal-organic framework material, iron terephthalate (MOF-235), J. Hazard. Mater. 185 (2011) 507-511. https://doi.org/10.1016/j.jhazmat.2010.09.035

[68] C. Brahmi, M. Benltifa, M. Ghali, F. Dumur, C. Simonnet-Jégat, V. Monnier, F. Morlet-Savary, L. Bousselmi, and J. Lalevée, Polyoxometalate s/polymer composites for the

photodegradation of bisphenol-A, J. Appl. Polym. Sci. 138, no. 34 (2021) 50864.
https://doi.org/10.1002/app.50864

[69] H. R. Ghalebi, S. Aber, and A. Karimi, Keggin type of cesium phosphomolybdate synthesized via solid-state reaction as an efficient catalyst for the photodegradation of a dye pollutant in aqueous phase, J. Mol. Catal. A: Chem. 415 (2016) 96-103. https://doi.org/10.1016/j.molcata.2016.01.031

[70] A. García, B. Rodríguez, M. Rosales, Y. M. Quintero, P. G. Saiz, A. Reizabal, S. Wuttke, L. Celaya-Azcoaga, A. Valverde, and R. F. D. Luis, A State-of-the-Art of Metal-Organic Frameworks for Chromium Photoreduction vs. Photocatalytic Water Remediation, Nanomater. 12, no. 23 (2022) 4263. https://doi.org/10.3390/nano12234263

[71] M. A. Basith, N. Yesmin, and R. Hossain, Low temperature synthesis of BiFeO 3 nanoparticles with enhanced magnetization and promising photocatalytic performance in dye degradation and hydrogen evolution, RSC Adv. 8 (2018) 29613-29627. https://doi.org/10.1039/C8RA04599B

[72] K. Guesh, C. AD Caiuby, A. Mayoral, M. Díaz-García, I. Díaz, and M. Sanchez-Sanchez, Sustainable preparation of MIL-100 (Fe) and its photocatalytic behavior in the degradation of methyl orange in water, Cryst. Growth Des. 17 4 (2017) 1806-1813. https://doi.org/10.1021/acs.cgd.6b01776

[73] N. M. Mahmoodi, J. Abdi, M. Oveisi, M. A. Asli, and M. Vossoughi, Metal-organic framework (MIL-100 (Fe)): Synthesis, detailed photocatalytic dye degradation ability in colored textile wastewater and recycling, Mater. Res. Bull. 100 (2018) 357-366. https://doi.org/10.1016/j.materresbull.2017.12.033

[74] N. M. Mahmoodi, and J. Abdi, Nanoporous metal-organic framework (MOF-199): Synthesis, characterization and photocatalytic degradation of Basic Blue 41, Microchem. J. 144 (2019) 436-442. https://doi.org/10.1016/j.microc.2018.09.033

[75] N. Liu, C. Jing, Z. Li, W. Huang, B. Gao, F. You, and X. Zhang, Effect of synthesis conditions on the photocatalytic degradation of Rhodamine B of MIL-53 (Fe), Mater. Lett. 237 (2019) 92-95. https://doi.org/10.1016/j.matlet.2018.11.079

Emerging Materials for Next Frontier Energy and Environment Applications   Materials Research Forum LLC
Materials Research Foundations 170 (2024) 41-59   https://doi.org/10.21741/9781644903292-3

Chapter 3

Recent advances in the electrode materials for electrocatalytic hydrogen evolution reactions

Thirugnanam Bavani[1,2], Sakthivel Vinith Kumar[3], Jagannathan Madhavan[3*]

[1] Department of Chemistry, Karpagam Academy of Higher Education, Coimbatore, 641 021, India

[2] Centre for Energy and Environment, Karpagam Academy of Higher Education, Coimbatore, 641 021, India

[3] Solar Energy Lab, Department of Chemistry, Thiruvalluvar University, Vellore, 632 115, India

* jagan.madhavan@tvu.edu.in

Abstract

The innovative design of electrodes has had a profound impact on the improvement of electrocatalysts for water splitting, paving the way for feasible energy production and hydrogen creation through direct water electrolysis. Electrochemical hydrogen evolution is pivotal for efficient, cost-effective hydrogen generation, yet attaining highly active catalysts with low overpotential (OVP) using affordable materials remains challenging. This chapter assesses diverse electrode materials' influence and delves into current HER process understanding. It also discusses general electrocatalyst design concepts and compiles recent advances in HER electrocatalyst development. Finally, the chapter offers perspectives on potential directions for future research in this field.

Keywords

Hydrogen Evolution Reaction, Water Splitting Reactions, Electrocatalysis, Green Energy

Contents

Recent advances in the electrode materials for electrocatalytic hydrogen evolution reactions 41

1.   Introduction 42

2.   General Principle of Hydrogen Evolution Reaction 43

    2.1   Experimental electrochemical activity of HER catalysts 43

3.   Various electrocatalytic materials for HER 45

    3.1   Transition metal based electrocatalysts 45

3.2   Biomass derived electrode materials for HER 47

3.3   Quantum dots electrocatalysts...49

3.4   Carbon supported electrocatalysts for HER 50

3.5   Noble-metal electrode materials for HER 50

3.6   MOF and LDH electrode materials for HER..51

Summary and future prospects **54**

References...**55**

1.   Introduction

Population and industry inflation at an unprecedented rate has escalated the world's energy requirements. One of the significant and most urgent challenges of the 21st century is energy production. On account of these, there is an emerging need to develop sustainable surrogates for petroleum-based fuels. This can be achieved by increasing the amount of renewable energy as it improves sustainability and addresses the growing prominence of serious environmental issues [1-3]. Subsequently hydrogen (H_2) is able to be created from a different of bases, including natural gas, biomass, water, etc., it has been touted as one of the promising options for the upcoming energy demands as a fuel that is environmentally friendly (Fig. 1). The creation and application of hydrogen energy derived from renewable resources has emerged as a major global development avenue, in view of great efficiency, abundant materials, and zero pollution [4-7]. Therefore, methane steam reforming and coal gasification are the commercial sources of H_2. However, these two approaches come with a high cost and pose environmental problems [8–11]. Research is being urged to concentrate on green and renewable energy sources due to the fact that splitting naturally occurring, abundant water produces H_2 energy [12]. Due to the straightforward production procedure, massive manufacturing capacity, economic viability, and environmental friendliness, electrocatalytic evolution of H_2 from water has been selected as an intriguing strategy for sustainable H_2 production among several H_2 production processes. The selection of catalyst influences the electrocatalytic efficiency of the hydrogen evolution reaction (HER); the most suitable catalyst possesses a binding energy for adsorbed H+ which is comparable to the reactant or products. For HER, Pt-based metals are typically employed as highly effective electrocatalysts. Still, Pt's limited availability and cost-effectiveness as a noble metal severely impede its commercialization. It sparked a lot of research to create high HER activity alternative electrocatalysts free of noble metals. However, there is lot of transition metal-based and metal-free, carbon supported, MOFs and LDHs and quantum dots are prevalent in the earth that perform admirably for HER and/or Oxygen Evolution Reaction (OER) [13-17].

This book chapter focuses on the enhancement of recently explored noble and non-noble electrocatalysts for HER. More particularly carbides, nitrides, oxides, sulphides, phosphides and selenides of transition metals have initiated an attention. The basic mechanisms for the HER are also presented. Finally, an outlook of the major areas that requires scientific investigation in this rapidly expanding field are delivered.

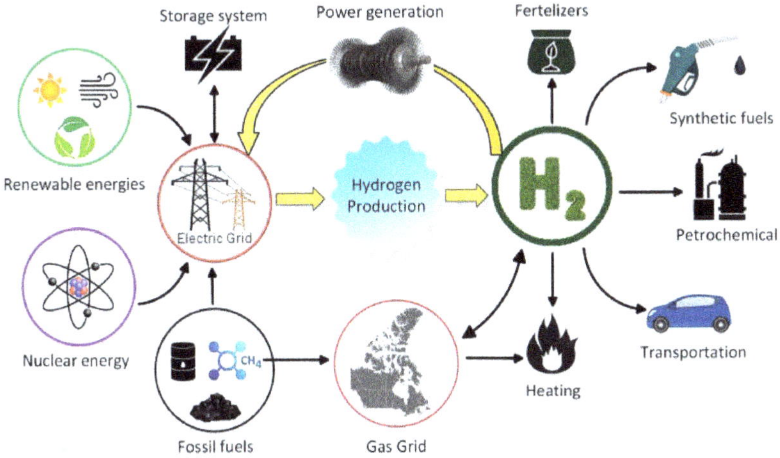

Emerging Materials for Next Frontier Energy and Environment Applications   Materials Research Forum LLC
Materials Research Foundations 170 (2024) 41-59   https://doi.org/10.21741/9781644903292-3

Fig. 1. Significance of hydrogen in participating various energy sectors. Copyright with permission from Ref. [7].

2.   General Principle of Hydrogen Evolution Reaction

The mechanistic step for the generation of H_2 is the proton reduction or reduction of water reaction taking place on the surface catalyst is the media of acid or alkaline electrolyte respectively. Before the reduction, there occurs a formation of adsorbed/desorbed intermediates (H^*). Subsequently, the reduction occurs by two mechanisms, firstly the Volmer-Heyrovsky mechanism, which entails the reaction of H^* and H_3O^+ (H_2O) by an electrochemical bonding and secondly, the Volmer-Tafel path by which the combination of two H^* takes place. (Fig. 2) [18-20]. The mechanism through which the HER is intuited can be determined by the experimental values of Tafel slope (TS), which demonstrates the rate-determining steps of Tafel, Heyrovsky and Volmer reactions.

2.1   Experimental electrochemical activity of HER catalysts

There are some primary parameters required to explicate the electrocatalytic activity of a particular electrocatalyst precisely which substantially include Tafel plot, stability, turnover frequency and faradic efficiency.

Total electrode activity

Generally, total electrode activity is first assessed by the execution of linear sweep voltammetry (LSV) or cyclic voltammetry (CV). Because, particularly for those carbon-containing catalysts, the capacitive current in the non-Faradaic zone could make up a fraction of the overall current we have recorded, provides us an approximate assessment from the CV or LSV results. A material's steady state currents at different voltages being applied was needed to be recorded at a period of at least 5 min, in order to more accurately establish its capacity for catalysis. Two special OVPs are often used in HER, to deliberately compare the sample activities. The "onset OVP" is one of them, which really has a wide connotation, and the other is relevant OVP, the potential of an electrocatalyst at a current density (CD) of 10 mA cm^{-2}

Tafel Plot and exchange current

The connection within various OVPs and steady-state CD's are explained by a Tafel plot. The TS and the exchange CD(j_0) are derived from the Tafel equation $\eta = a + b \log j$, whereby, b denotes the TS. The potential (where cathode and anode currents are equal) of a reaction is the exchange CD. The larger the exchange CD on the electrocatalyst surface, the higher the electrolytic activity.

Stability

Stability is a crucial factor which demonstrates the real activity of an electrocatalyst over an extended duration. In general, the stability of the HER catalyst can be discovered via two common ways of characterization. The first is to assess the current variation with time (I-t curve) and the other is by performing large cycles of CV or LSV (more than 5000 cycles).

Turnover Frequency

Turnover frequency (TOF), the number of the product molecules that a catalyst converts per catalytic area in a unit of time, where the intrinsic action of the catalytic unit is exhibited.

Faradaic efficiency

For the electrochemical HER, Faradaic efficacy is designated as the ratio of the experimentally perceived H_2 quantity to the theoretical H_2 quantity that, assuming a 100% Faradaic yield, could be computed from the CD.

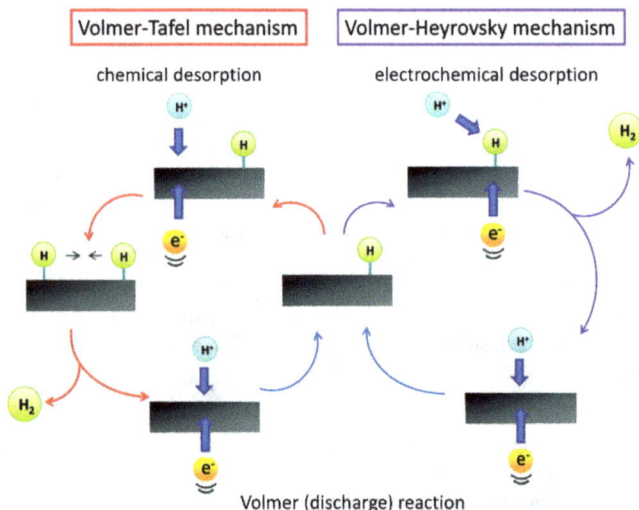

Fig. 2. The mechanism for the evolution of hydrogen on the surface of an electrode in acidic solution. Copyright with permission from Ref. [19]

3. Various electrocatalytic materials for HER

3.1 Transition metal based electrocatalysts

Research on sustainable and renewable energy has recently evoked an inordinate exchange of interest among the researchers towards appropriate designing of non-precious metal electrocatalysts for greatly effective HER in alkaline media. Yang et al. reported a procedure with two stages that started with the hydrothermal preparation method for Ni-Co molybdate nanosheet (NS) assembly as the precursor and arrived with vapour phase sulfurization to carry out in-situ transformation [21]. It was used to create a vertically supported and interlinked $NiS_2/CoS_2/MoS_2$ NS on Ni foam (Fig 3). The structure like a honeycomb of the Ni-Co molybdate NS assembly was found to be well retained during the sulfurization procedure based on the characterizations obtained by SEM and TEM. The $NiS_2/CoS_2/MoS_2$ NS provided many heterointerfaces and a huge number of imperiled activated sites, which contributed to the improved alkaline HER efficacy. Using the well-exposed multiple heterointerfaces, the ultrathin $NiS_2/CoS_2/MoS_2$ NS displayed an exceptional HER efficiency with 112 mV of OVP at 10 mA cm^{-2} and a 59 mV dec^{-1} of TS of, as displayed through electrochemical investigation, when compared to CoS_2/MoS_2 (124 mV and 75 mV dec^{-1}) and NiS_2/MoS_2 (155 mV and 89 mV dec^{-1}) materials. A study of heterostructure consisting of CoS_2 and CoSe on carbon matrix ($CoS_2/CoSe@C$) by Karuppasamy et al. was reported [22]. The sulphur and selenium content of the subsequent $CoS_2/CoSe@C$ electrocatalyst was rationally regulated to improve the HER efficiency. The produced electrocatalyst delivers 164 mV of OVP at 10 mA cm^{-2} CD with a 42 mV dec^{-1} of TS value, due to the appropriate synergy among the CoS_2 and CoSe. Furthermore, improved conductivity was also achieved via the interrelated structure in the carbon matrix of $CoS_2/CoSe@C$.

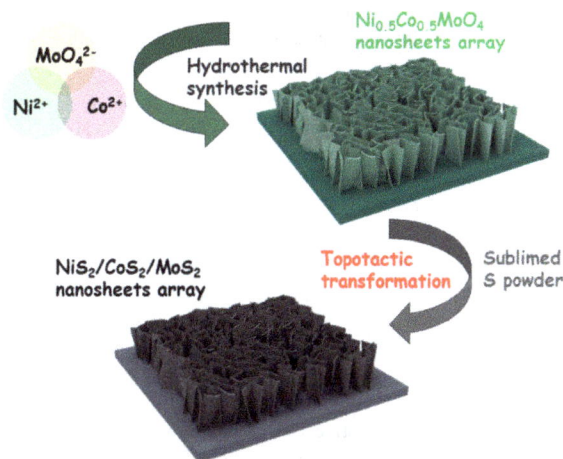

Fig. 3. Schematic demonstration of the two-step approach of assembling honeycomb-like mesoporous $NiS_2/CoS_2/MoS_2$ NSAs. Copyright with permission from Ref. [21].

Using heterojunction engineering, Sun et al. constructed a unique flower-like WS_2/WSe_2 heterostructure [23]. Applying a hierarchical construction, this multiscale morphological modulation approach exploits the structural and compositional attributes of the WS_2/WSe_2 heterojunction to substantially improve the contact of many of the electrochemical active areas through regulating the electronic structure. The WS_2/WSe_2 heterostructure exhibited excellent HER properties, as predicted, including a small slope of TS of 74.08 mV dec^{-1} and 121 mV of OVP at 10 mA cm^{-2}. It additionally demonstrated a notable endurance. Similarly, Yang et al. stated on the synthesis a set of 3D-porous iron foam (IF) self-supported FeS_2–MoS_2 bimetallic hybrid compounds [24]. These compounds were referred as FeS_2–MoS_2@IFx, where x = 100, 200, 300, and 400 mg of $(NH_4)_6Mo_7O_{24}$·$4H_2O$ respectively. It was significant to note that polyoxomolybdates $(NH_4)_6Mo_7O_{24}$·$4H_2O$, a source of Mo, and iron ferrocyanide that served as a slow-releasing iron source which was sulfurized on addition of thiourea to generate self-supported FeS_2–MoS_2 on IF (also known as FeS_2–MoS_2@IF200), as an effective electrocatalyst. Further characterizations revealed that FeS_2-MoS_2 resembled flower clusters, grown on the 3D- Fe-foam framework and was made up of staggered, linked NS with mesoporous structures.

Ling et al. synthesized N-doped CoP_2 incorporated on porous carbon cloth (PCC) for effective HER in an alkaline electrolyte. The as-prepared CoP_2@PCC necessitated only 64 mV of OVP to generate an average CD of 10 mA cm^{-2}. It exhibited low TS of 47.4 mV dec^{-1} and exhibited exceptional stability for 24 h at a constant CD of 10 mA cm^{-2} in 1 M KOH. The Faradaic efficiency exposed by CoP_2@PCC was ~100% (Fig. 4) [25]. Also, Ni et al. reported on an exceptional performance and endurance for HER in a neutral medium exhibited by nanocomposite nitrides composed of deficient HER electrocatalysts, specifically Ni_3N and VN. The Ni_3N-VN composite is related to the most advanced Pt/C electrocatalyst, needing only 85 mV OVP to achieve a standard 10 mA cm^{-2} of CD in 1.0 M saline solution with phosphate buffer. When compared to Ni_3N and VN, the HER kinetics are markedly enhanced, exhibiting CD 6–17 fold greater and lowered TS [26].

Ying and co-workers, reported on the development of a durable W_2C hollow microsphere (HS) electrocatalyst with enduring electrocatalytic performance towards HER. This catalyst is made by carburizing tungsten oxides at 700 °C with CH_4/H_2 stream and this electrocatalyst required OVP of 264 and 153 mV to provide 100 and 10 mA cm^{-2}, individually, corresponding to the hollow structures that were advantageous for both interfacial migration of charges and hydrogen molecular release. In the meantime, the W_2C-HS electrocatalyst showed an imperceptible deterioration later 20,000 cycles which indicated an exceptional durability [27].

Yi et al. innovated nanohybrid referred to as P-Mo_2C/Ti_3C_2@NC that contained molybdenum carbide nanodots loaded with phosphorus and anchored on Ti_3C_2 flakes encased in N-doped carbon. This combination of materials was believed to function as a very active non-precious metal electrocatalyst for HER. The P-doped Mo_2C nanodots exhibited an excellent dispersion and limited growth due to the conductive Ti_3C_2 based intrinsic anchoring sites. This resulted in an optimized interfacial interaction with the conductive Ti_3C_2 medium and the P -doped Mo_2C nanodots. In the meantime, the Ti_3C_2 flakes were also stabilized against spontaneous oxidation by porous carbon shells doped with nitrogen. Therefore, an adequate addition of a Ti_3C_2 material and the cooperative catalytic interfaces within P-Mo_2C nanodots and Ti_3C_2 encased by N-doped porous carbon gave an overall boost in the electrical conductivity and reactive site exposure. Hence, the P-Mo_2C/Ti_3C_2@NC hybrid catalyst exceeded the Ti_3C_2@NC, P-Mo_2C/rGO@NC, P-Mo_2C/NC and P-Mo_2C@NC nanocomposites of HER activity. It presented 177 mV at 10 mA cm^{-2} of OVP, 57.3

Emerging Materials for Next Frontier Energy and Environment Applications   Materials Research Forum LLC
Materials Research Foundations 170 (2024) 41-59   https://doi.org/10.21741/9781644903292-3

mV dec^{-1} of fast reaction kinetics, and stable behaviour in the acidic electrolyte over 60 h. This study encouraged the investigation of an emerging field of MXene-based nanocomposites for applications in green energy, thus creating a path for the creation of sophisticated MXenes-based electrocatalysts for HER [28].

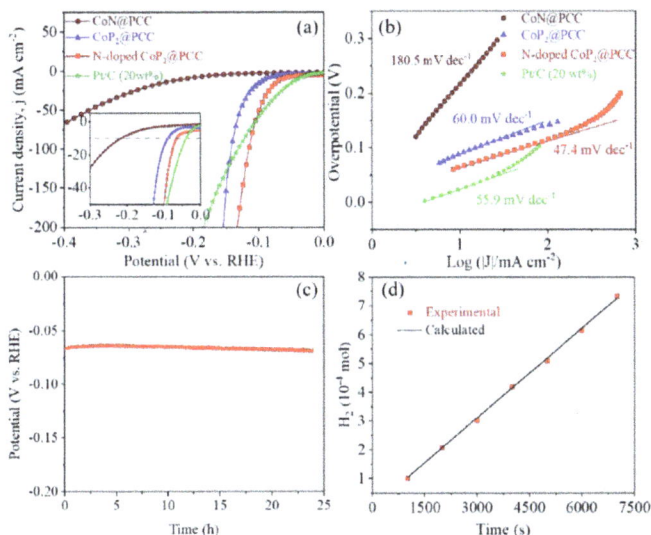

Fig. 4. (a) LSV plots and (b) Tafel plots of Pt/C (20 wt%), CoN@PCC, CoP$_2$@PCC, and N-doped CoP$_2$@PCC for HER in 1 M KOH with a scan rate of 5 mV s^{-1}. The inset of (a) illustrates the LSV plots in the CD range of 0 to -50 mA cm^{-2}. (c) Chronopotential curve of N-doped CoP$_2$@PCC at the constant CD of 10 mA/cm^2 for 24 h. (d) The molar volume of hydrogen experimentally calculated Vs. theoretically computed for N-doped CoP$_2$@PCC at a current of 20 mA. Copyright with permission from Ref. [25].

3.2   Biomass derived electrode materials for HER

In the past few decades, the resources of carbon from different biomass have played a crucial role in preparing carbon doped electrode materials. Fakayode et al. reported on Mo$_2$C/C electrocatalyst derived from biomass [29]. A facile procedure with two stages and simple technologies were used to generate biomass derived carbon from watermelon peels for HER application. The prepared Mo$_2$C/C electrocatalysts had distinctive pore diameters and surface areas at a temperature of 800 °C. 1.0 M of KOH with a geometric OVP of 133 mV gave a CD of 10 mA cm^{-2} and an additional 300 h long-term durability process was also observed. As a result, this study developed a novel method for creating impeccable HER electrocatalyst from biomass waste.

Min et al. designed a self-supporting H$_2$ evolution cathode made of porous carbon (PPDC) membrane derived from pomelo peel (PP) and imbedded with Co nanoparticles (Co@PPDC) which was generated by directly carbonizing the Co-adsorbed PP (Co^{2+}-PP) [30]. The self-supported Co@PPDC electrode exhibited exceptional electrocatalytic activity for HER in 1.0 M KOH solution, with OVPs of 264 and 154 mV at CD's of 100 and 10 mA cm^{-2}, individually. With around 100% Faradaic efficiency (FE) and 1.56 mmol h^{-1} of H_2 production rate, the Co@PPDC electrode also displayed an tremendous recyclability for 2000 runs and a 100 mA cm^{-2} of CD at a constant OVP of 265 mV over 12 h.

Sekar et al. created a corrugated graphene NS (RH-CG) using biomass obtained from rice husks using the KOH activation process. Both a greater electrical conductivity and specific surface area was observed in the 700 °C-activated RH-CG NS. When used as a HER electrocatalyst in 0.5 M H_2SO_4, the RH-CG NS produced an outstanding HER activity with a minimal OVP (9 mV at 10 mA cm^{-2}) and TS (31 mV dec^{-1}). The outcomes presented a novel approach of an outstanding electrocatalyst composed of biomass for highly operative hydrogen production [31].

Fu et al. investigated a KOH activation process to produce biomass carbon materials using walnut shell (WS) for HER electrocatalysts [32]. The structural characteristics of the surface of the walnut shell-derived carbon series, like surface area, pore structure, and morphological were reported to be affected by the activation temperature. The hierarchical porous carbon material, that underwent the activation at 600 °C, contained micropores with a specific area of 1037.31 m^2 g^{-1} and 0.51 cm^3 g^{-1} of volume, which resulted in good electrochemical activity of HER. In particular, the carbon material produced from walnut shells activated at 600 °C had shown exceptional stability after a long-term cycling, 6.00 mV of lower onset potential and TS of 69.76 mV dec^{-1} (Fig.5).

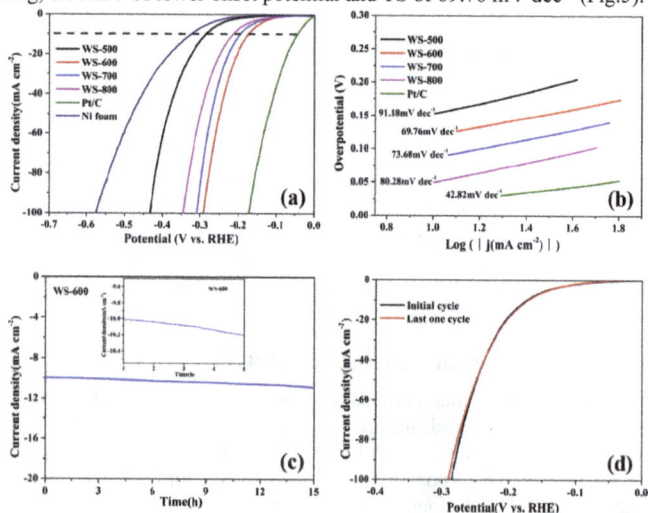

Fig. 5 (a) LSV plots and (b) Tafel plots of WS-X materials and Pt/C for HER in 1.0 M KOH. (c) The time-dependent current density of WS-600 electrode at the potential of 0.16 V vs. RHE. (d) Polarization plots of WS-600 measured before and after 15 h potential sweeps in 1.0 M KOH. Copyright with permission from Ref. [32].

3.3   Quantum dots electrocatalysts

Fu et al.'s hydrothermal-electrodeposition technique was used to create a Co_xP QDs-MoS_2-Ni_3S_2-NS@NF with electrocatalytic overall water splitting [33]. On the surface of the NF, the $MoS_2@Ni_3S_2$/NF was created using a hydrothermal approach, and the Co_xP QDs were later applied using an electrodeposition technique to the MoS_2 NS surface. As shown, the synthesized Co_xP QDs-$MoS_2@Ni_3S_2$/NF possessed 54 mV of OVPs at 10 mA cm^{-2} and 207 mV (OER) at 100 mA cm^{-2} in 1.0 M KOH, and demonstrated an adequate overall water splitting efficiency, with an OVP of 1.39 V (vs RHE) at 10 mA cm^{-2}. This was mostly attributed to the MoS_2 NS and Co_xP QDs which effectively boosted the HER/OER activity.

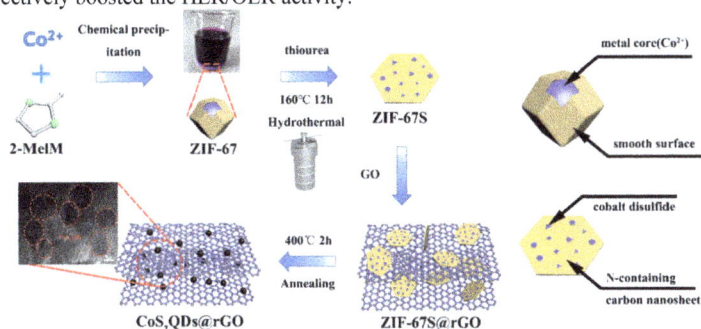

Fig.6. Schematic diagram of the synthesis of $CoS_2QDs@rGO$. Copyright with permission from Ref. [36].

Liu et al. expressed how the Ti_2CT_x MQDs spontaneously evolved surface groups (x: the amount of O atoms present), i.e., the -Cl functional groups were substituted for O-terminated whereas the cathode reaction occurs [34] (Fig. 6). Due to this mechanism, HER had 0.26 eV of low Gibbs free energy. The stable Ti_2CO_x/Cu_2O/Cu foam systems had displayed a good operative durability over 165 h at a fixed CD of 10 mA cm^{-2}, besides a small OVP of 175 mV at 10 mA cm^{-2} in 1 M KOH.

Huo and co-workers. innovated a distinct 0D/2D heterojunctions of uniform $(Mo_2C)_x$-$(WC)_{1-x}$-QDs (3-5 nm)/N-doped graphene (NG) NS with a 2D- layered structure that were created by nanocasting, utilizing the KIT-6/graphene (G) using template [35]. The study showed the possibility to create $(Mo_2C)_x$-$(WC)_{1-x}$-QDs ($0 < x < 1$)/NG nanocomposites with various Mo/W molar proportions by manipulating the molar proportion of the W and Mo precursors. These hybrids were found to be more active in the HER than the pure Mo_2C/NG and WC/NG nanocomposite. The higher performance of the HER was suggested to be ascribed to the displacement of the valence electrons of the W and Mo elements, nitrogen coordinating sites, and well scattered Mo_2C-WC nanocrystals, in addition to the robust connection between NG and Mo_2C-WC nanocrystals. It is fascinating to note that the ideal electrocatalyst, $(Mo_2C)_{0.34}$-$(WC)_{0.32}$/NG displayed small OVPs (100 and 93 mV) to attain a cathodic CD of 10 mA cm^{-2}, smaller TS (53 and 53 mV dec^{-1}), and large interchange CD's in acidic and alkaline media, correspondingly.

Jiang et al. using ZIF-67 as a precursor and developed 3D-multilayer rGO supported CoS_2 quantum dots (CoS_2QDs@rGO) using an a single-step calcinations approach and a two-step hydrothermal process (Fig. 6) [36]. The electrocatalyst was reported to be highly effective for HER, in sites of remarkably low TS of 78 mV dec^{-1} and substantial stability it was furthermore produced with the strongest accessibility of edge active sites. CoS_2QDs@rGO had shown an outstanding electrochemical performance in alkaline solutions in the interim. The DFT calculations have shown that the revelation of active sites and changes in the sample morphology had significantly enhanced the HER activity.

3.4   Carbon supported electrocatalysts for HER

`Prem et al. created a Ni_2P supported carbon nanotubes (CNTs) in both alkaline and acidic environments, respectively. Ni_2P/CNT has been used as an effective electrocatalysts for OER and HER. On 10 mA cm^{-2}, the electrocatalyst had demonstrated lower OVPs of 360 and 137 mV for the OER and HER, correspondingly. Reports on lower TS, increased stability, and increased electrochemical active surface area were also included in the study [37].

A simple hydrothermal one-step synthesis of Co-doped $Ni_{0.85}Se$ nanoparticles anchored on RGO (Co-$Ni_{0.85}Se$/RGO) was reported by Xu et al [38]. They created hybrid compounds successfully prevented the nanoparticle aggregation and resulted in an improved intrinsic conductivity, which exposed more active sites and accelerated the electron migration. Owing to all of these advantageous impacts, Co-$Ni_{0.85}Se$/RGO was found to be capable of efficient HER catalyst in both basic and acidic electrolytes, in terms of small OVPs of 128 and 148 mV at 10 mA cm^{-2} of CD, in 1.0 M KOH and 0.5 M H_2SO_4 individually.

In a study of one-step sonochemical process, Renuka et al. created the Cu_2ZnSnS_4/MoS_2-rGO heterostructure electrocatalyst [39]. Having an onset potential as low as 50 mV vs. RHE at 10 mA cm^{-2}, a TS of 68 mV dec^{-1}, and 962 mA cm^{-2} of exchange CD, the Cu_2ZnSnS_4/MoS_2-rGO exhibited an exceptional HER activity. Being coupled with MoS_2-rGO, the CZTS has confirmed a synergistic impact that created an outstanding, long-lasting H_2 evolution electrocatalytic system.

Moreover, the halide Perovskite electrode materials for HER was reported by Sarmad et al., on Pr doped in the $SrTiFeO_{6-\delta}$ on A- and B-site to ascertain its activity as an electrocatalyst [40]. Herein, both A and B-sites doped Pr in $SrTiFeO_{6-\delta}$ (STFP01) double perovskite structure presented tremendous outcomes with an OVP of 182.4 mV at 5 mV s^{-1} of scan rate of and 77 mV dec^{-1} TS. The sponge-like form of Pr-doped $SrTiFeO_{6-\delta}$ had displayed a remarkable potential to be a sustainable electrocatalyst.

3.5   Noble-metal electrode materials for HER

Cui et al.'s study on hierarchical Pt-MXene-SWCNTs heterostructures for HER catalysts is proposed below. Remarkably effective nano/atom-scale metallic Pt was arrested on $Ti_3C_2T_x$ MXene flakes (MXene@Pt) within the heterostructure and was coupled via a network of conductive SWCNTs [41]. Filtration was implied to create a hierarchical heterostructure from a mixed colloidal suspension of MXene@Pt and SWCNTs. The MXene@Pt suspension of colloids was synthesized through simply reducing the Pt cations to metallic Pt without the use of added reductants or post-treatments, thereby taking the advantage of reducibility and hydrophilicity characteristics of MXene. The membrane-based, so-fabricated hierarchical HER catalysts were found to exhibit excellent stability over 800 h of process, a 230 mA cm^{-2} of high-volume CD up to at a potential difference of 50 mV from the RHE, and 62 mV of low OVP at a CD of 10 mA cm^{-2}.

Zhu et al. studied the HER activity of Pt catalysts with various structures with single atoms (SAs), nanoparticles and clusters, that have been successfully and cost-effectively anchored on VS$_2$ NS [42]. With both clusters and Pt SAs, the effective Pt-ornamented VS$_2$ catalyst has produced an OVP of 77 mV at 10 mA cm^{-2}, which was comparable to Pt/C's (48 mV) OVP.

Ruqia et al. composited Pt-loaded CoP connected through ethylene-glycol proton concentration in which hydrogen spillover overcame by using specific ethylene glycol ligand conditions [43]. The catalytic performance was greatly enhanced at a modest Pt loading of 1.5 wt%, and the TS were drastically reduced, going from 104.6 mV dec^{-1} of CoP to 42.5 mV dec^{-1}. These composites with 1.5 wt% Pt stacking were found to outperform the viable Pt/C and other hydrogen spillover electrocatalysts for HER, delivering a small OVP of 21 mV at 10 mA cm^{-2} and a measure high noble-metal operation efficiency.

Yao et al.'s study on controlled deposition of sub-monolayer Pt on an intermetallic Pd$_3$Pb nanoplate (AL-Pt/Pd$_3$Pb) is discussed below. In this study, the active surface of Pt layer's atomic efficiency and electrical structure were extensively optimized which significantly improved the acidic HER [44]. With an OVP of only 13.8 mV at 10 mA cm^{-2} and a great mass action of 7834 A g^{-1}$_{Pd+Pt}$ at -0.05 V, AL-Pt/Pd$_3$Pb displayed an exceptional HER performance, far above than that of viable Pt/C (30 mV, 1486 A g^{-1}$_{Pt}$). Further AL-Pt/Pd$_3$Pb possessed the additional characteristics of excellent durability and stability.

3.6   MOF and LDH electrode materials for HER

Cheng et al. devised an intuitive approach to create an attractive electrocatalyst made of synergetic iron hydr(oxy)oxide [Fe(OH)$_x$] nanoboxes and a nanoscale conductive copper-based MOF (Cu-MOF) layer [45]. Thus produced Fe(OH)x@Cu-MOF nanoboxes demonstrated an enhanced performance and the stability during the electrocatalytic HER owing to the significantly accessible active sites, improved charge migration, and strong hollow nanostructure. To be more precise, TS of 76 mV dec^{-1} and a CD of 10 mA cm^{-2} required OVP of 112 mV.

Hao co-worker. fabricated the binary transition metal phosphide (Co$_x$Fe$_{1-x}$P) nanocubes (NC) that have various Co and Fe ratios by a phosphidation procedure via MOFs as templates [46]. Following phosphidation, MOF templates provided NC structural characteristics and the creation of crystals while maintaining the defined structure were made possible at the right phosphidation temperature. The valence electrons in Co$_x$Fe$_{1-x}$P were redistributed as a result of the insertion of a binary transition metal. These modifications suggested that the P and Fe atoms were in anionic states that acted as active sites to donate electrons of Co$_x$Fe$_{1-x}$P NC. These distinctive properties supported the HER activity. Surprisingly, despite a low OVP of 92 and 72 mV in alkaline and acidic conditions respectively. The Co$_{0.59}$Fe$_{0.41}$P NC produced at 450 °C provided a CD of 10 mA cm^{-2}. Moreover, Co$_{0.59}$Fe$_{0.41}$P NC also displayed a negligible TS value of 52 mV dec^{-1} in acidic and 72 mV dec^{-1} in alkaline condition. These NC were also reported to exhibit outstanding durability in both alkaline and acidic environments. According to Hao et al., their highly active HER catalyst that was created using the technique of binary transition metal MOF templates has opening up the new possibilities for creating much more superior electrocatalysts [46]. Similarly, Do et al. too reported on Ni-based MOF that was produced in-situ on carbon fibre (NiSe$_2$/C/CF) over pyrolysis and selenization methods. NiSe$_2$/C/CF seemed to perform better on the HER test than Ni/C/CF and Ni-MOF-74/CF (Fig.7). Mainly, the NiSe$_2$/C/CF electrode was exhibiting an exceptional stability with only a little decline after working for 12 h with a TS of 74.1 mV dec^{-1}, and an OVP of 209 mV [47]. According to their reports, the emergent effects of NiSe$_2$ nanoparticles and 3D- conductive substrate CF has facilitated the active compounds availability and migration of electrons throughout the electrocatalytic route which were primarily responsible for the remarkable HER catalytic efficiency of NiSe$_2$/C/CF.

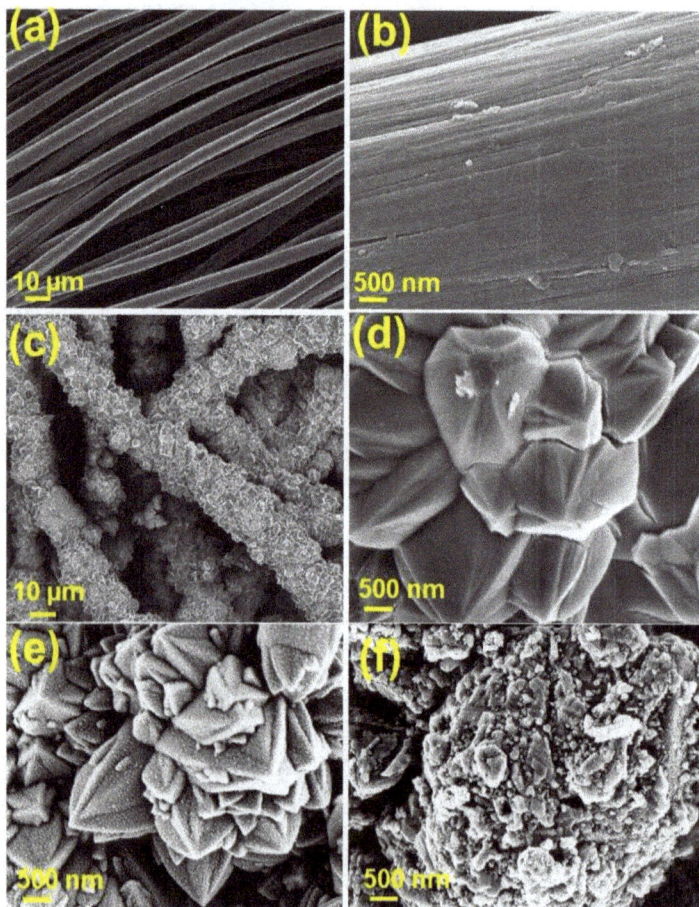

Fig.7. SEM images of (a, b) CF, (c,d) Ni-MOF-74/CF, (e) Ni/C/CF and (f) NiSe$_2$/C/CF. Copyright with permission from Ref. [47].

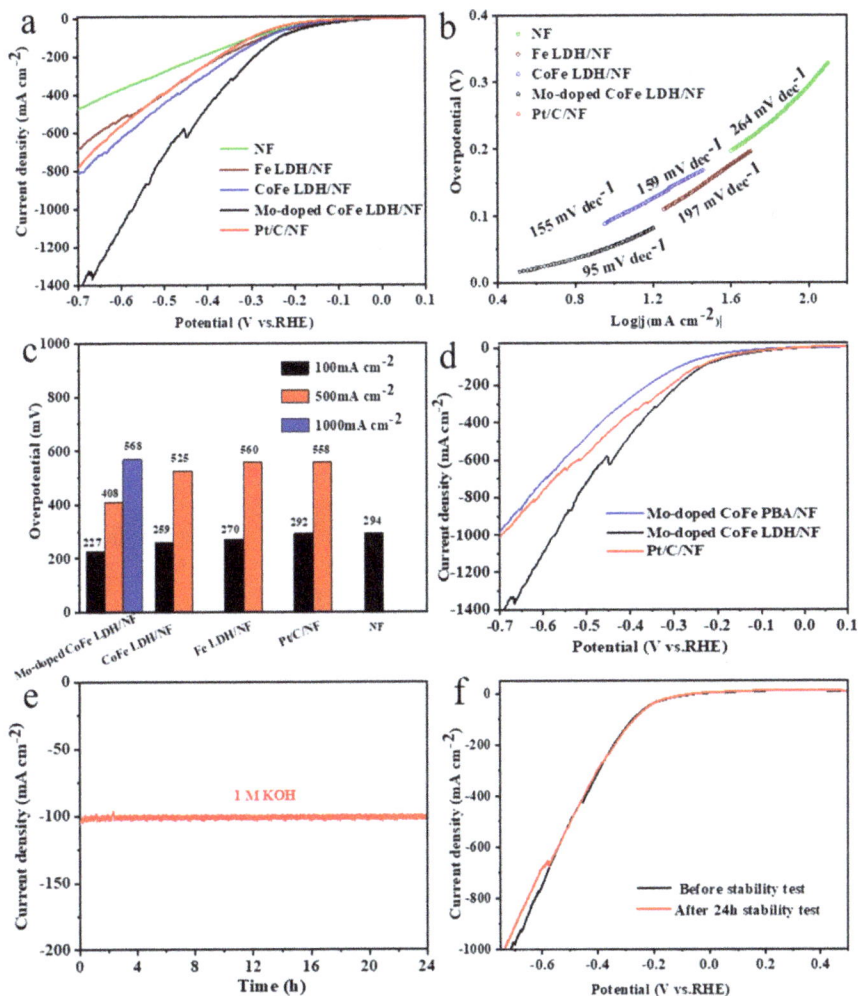

Fig. 8. a) HER polarization plot and (b) corresponding Tafel plots of NF, Fe LDH/NF, CoFe LDH/NF, Mo-doped CoFe LDH/NF and Pt/C/NF; (c) OVPs at standard CD's of several catalysts; (d) Polarization curves of Mo-doped CoFe PBA/NF, Mo-doped CoFe LDH/NF and Pt/C/NF; (e) The durability studying of Mo-doped CoFe LDH/NF at the CD's of 100 mA cm⁻²; (f) LSV plots of Mo-doped CoFe LDH/NF after the durability assessment 24 h. Copyright with permission from Ref. [48].

The Mo-doped Prussian blue NC of CoFe LDH on Ni foam was reported by Zhao et al. for effective water splitting [48]. The electrocatalyst was reported to exhibit remarkable HER, due to the highly stabilized active sites. It showed 227 mV OVP @ 100 mA cm^{-2} in alkaline medium with excellent stability. Sang et al. too designed a CoNi-LDH/Co@NC that was presented with attractive electrocatalytic capabilities for HER in alkaline conditions. The material was developed by growing CoNi-LDH on ZIF-derived Co nanoparticles and N-doped carbon framework (Fig. 8). As per the results, CoNi-LDH/Co@NC only took 187 mV OVP when applied as HER electrocatalysts at a CD of 100 mA cm^{-2}. Also, Yang et al.'s stated the crystalline layered double hydroxide (LDH) becomes activated as an incredibly effective HER catalyst through the amorphization caused by B-doping [49]. As reported, the durability of the amorphous B-incorporated NiCo-LDH supported on Ni-foam than the commercial Pt was reflected in terms of high CD's at low OVPs (i.e.) 100, 500, 1000 mA cm^{-2} at 151, 286 and 381 mV correspondingly (Fig. 9).

The ion exchange and seawater corrosion were employed to create nickel–iron LDH and Ni_3S_2 heterostructured nanoflower bifunctional electrocatalysts (NiFe LDH-Ni_3S_2) by Zhang et al. [50]. Seawater corrosion time adjusted the catalyst's nanoflower structure, which resembled a desert rose. This peculiar heterostructure was reported to contain additional redox reaction centers, and also possessed improved anti-corrosion properties thus allowed it to maintain the high electrocatalytic activity in salt water while efficiently facilitating the mass diffusion and charge transport. For HER in 1 M KOH seawater, the NiFe LDH-Ni_3S_2 electrode's measured OVPs were as low as 257 mV.

Summary and future prospects

A vital approach to generate hydrogen to address the energy crisis is through electrolysis of water. In this context, major developments and advancements towards the designing and optimization of bifunctional electrocatalysts were reported in the last few decades. Reports on the compilation of the non-noble metal-based catalysts like oxides, metals, hydroxides, phosphides, chalcogenides, borides, carbides, nitrides, and so on, are discussed in this chapter. Of them, a few studies that have showed similar or even better results than the benchmark noble-metal based catalysts in terms of water splitting capabilities are quite encouraging in this field of research. Based on these outcomes, the following are most important optimizing ways that impacts on the outcomes, intrinsic actions and active sites: To rise the number of active centers, better morphology designing, to rise the conductivity of electricity, to accelerate up the motion of ions and electrons, implementation of multi-metal and defect engineering techniques to modify the electronic arrangement and states, employing heterostructures/interfaces to increase water splitting efficiency. To conclude, we suggest that combining hybrid approaches seemed to be more efficient way in achieving a balance between OER and HER applications.

With appreciable advancements in electrocatalysts made so far for general water splitting, several problems to be resolved for plausible HER applications.

While DFT theory assists investigators to comprehend stages in reactions, the intricate structure of composite catalysts still remains as a challenge to find the actual active sites in the catalysts.

Extended durability. The overall splitting of water is frequently carried out in an acidic or alkaline aqueous media, that possibly generates catalyst surface polarization. Maintaining high activity and stability in the electrolyte requires the protecting of catalysts from corrosion in order to execute the successful transition of academic research to commercial manufacturing. Hence, reasonable techniques addressing this crucial step can definitely prove beneficial in reducing electrode polarization.

Though several efficient water splitting catalysts for HER and OER have been developed and constructed to date, there is still a tremendous space for advancement outside of the research in terms of commercializing the developed catalysts that are extremely effective and cost-effective.

References

[1] A. Muzammil, R. Haider, W. Wei, Y. Wan, M. Ishaq, M. Zahid, W. Yaseen, X. Yuan, Emerging transition metal and carbon nanomaterial hybrids as electrocatalysts for water splitting: a brief review, Mater. Horiz., 10 (8) (2023) 2764-2799. https://doi.org/10.1039/D3MH00335C

[2] A. Lasia, Mechanism and kinetics of the hydrogen evolution reaction, Int. J. Hydrog. Energy, 44 (36) (2019) 19484-19518. https://doi.org/10.1016/j.ijhydene.2019.05.183

[3] G. Zhao, K. Rui, S.X. Dou, W. Sun, Heterostructures for electrochemical hydrogen evolution reaction: a review, Adv. Funct.l Mater., 28 (43) (2018) 1803291. https://doi.org/10.1002/adfm.201803291

[4] R.N. Iyer, H.W. Pickering, Mechanism and kinetics of electrochemical hydrogen entry and degradation of metallic systems, Ann. Rev. Mater. Sci., 20 (1) (1990) 299-338. https://doi.org/10.1146/annurev.ms.20.080190.001503

[5] X. Zou, Y. Zhang, Noble metal-free hydrogen evolution catalysts for water splitting, Chem. Soc. Rev., 44 (15) (2015) 5148-5180. https://doi.org/10.1039/C4CS00448E

[6] A.P. Murthy, J. Theerthagiri, J. Madhavan, Insights on Tafel constant in the analysis of hydrogen evolution reaction, J. Phys. Chem. C, 122 (42) (2018) 23943-23949. https://doi.org/10.1021/acs.jpcc.8b07763

[7] S. Farid, I. Dincer, A review and comparative evaluation of thermochemical water splitting cycles for hydrogen production. Energy Convers.Manag. 205 (2020) 112182. https://doi.org/10.1016/j.enconman.2019.112182

[8] B. You, Y. Sun, Innovative strategies for electrocatalytic water splitting, Acc. Chem. Res., 51 (7) (2018) 1571-1580. https://doi.org/10.1021/acs.accounts.8b00002

[9] Z.P. Ifkovits, J.M. Evans, M.C. Meier, K.M. Papadantonakis, N.S. Lewis, Decoupled electrochemical water-splitting systems: a review and perspective, Energy Environ. Sci., 14 (9) (2021) 4740-4759. https://doi.org/10.1039/D1EE01226F

[10] Y. Luo, Z. Zhang, M. Chhowalla, B. Liu, Recent advances in design of electrocatalysts for high-current-density water splitting, Adv. Mater., 34 (16) (2022) 2108133. https://doi.org/10.1002/adma.202108133

[11] M. Chatenet, B.G. Pollet, D.R. Dekel, F. Dionigi, J. Deseure, P. Millet, R.D. Braatz, M.Z. Bazant, M. Eikerling, I. Staffell, P. Balcombe, Water electrolysis: from textbook knowledge to the latest scientific strategies and industrial developments, Chem. Soc. Rev., 51 (11) (2022) 4583-4762. https://doi.org/10.1039/D0CS01079K

[12] H. Sun, X. Xu, H. Kim, W. Jung, W. Zhou, Z. Shao, Electrochemical water splitting: Bridging the gaps between fundamental research and industrial applications, Energy Environ. Mater., 6 (5) (2023) e12441. https://doi.org/10.1002/eem2.12441

[13] A.M. Ramírez, S. Heidari, A. Vergara, M.V. Aguilera, P. Preuss, M.B. Camarada, A. Fischer, Rhenium-based electrocatalysts for water splitting, ACS Mater. Au, 3 (3) (2023) 177-200. https://doi.org/10.1021/acsmaterialsau.2c00077

[14] A. Hayat, M. Sohail, H. Ali, T.A. Taha, H.I.A. Qazi, N. Ur Rahman, Z. Ajmal, A. Kalam, A.G. Al-Sehemi, S. Wageh, M.A. Amin, Recent Advances and Future Perspectives of Metal-Based Electrocatalysts for Overall Electrochemical Water Splitting, Chem. Record, 23 (2) (2023) e202200149. https://doi.org/10.1002/tcr.202200149

[15] A. Raveendran, M. Chandran, R. Dhanusuraman, A comprehensive review on the electrochemical parameters and recent material development of electrochemical water splitting electrocatalysts, RSC Adv., 13 (6) (2023) 3843-3876. https://doi.org/10.1039/D2RA07642J

[16] M.B. Wazir, M. Daud, S. Safeer, F. Almarzooqi, A. Qurashi, Review on 2D molybdenum diselenide (MoSe2) and its hybrids for green hydrogen (H2) generation applications, ACS Omega, 7 (20) (2022) 16856-16865. https://doi.org/10.1021/acsomega.2c00330

[17] T. Altalhi, S.M. Adnan, M.A. Amin (Eds.), Materials for Hydrogen Production, Conversion, and Storage, John Wiley & Sons, 2023.

[18] S. Wang, A. Lu, C.J. Zhong, Hydrogen production from water electrolysis: role of catalysts, Nano Converg., 8 (2021) 1-23. https://doi.org/10.1186/s40580-020-00251-6

[19] J. Su, J. Zhou, L. Wang, C. Liu, Y. Chen, Synthesis and application of transition metal phosphides as electrocatalyst for water splitting, Sci. Bull., 62 (9) (2017) 633-644. https://doi.org/10.1016/j.scib.2016.12.011

[20] A. Li, Y. Sun, T. Yao, H. Han, Earth-abundant transition-metal-based electrocatalysts for water electrolysis to produce renewable hydrogen, Chem. A Eur. J., 24 (69) (2018) 18334-18355. https://doi.org/10.1002/chem.201803749

[21] Y. Zhang, M. Shi, C. Wang, Y. Zhu, N. Li, X. Pu, A. Yu, J. Zhai, Vertically aligned NiS2/CoS2/MoS2 nanosheet array as an efficient and low-cost electrocatalyst for hydrogen evolution reaction in alkaline media, Sci. Bul., 65 (5) (2020) 359-366. https://doi.org/10.1016/j.scib.2019.12.003

[22] K. Karuppasamy, R. Bose, V.R. Jothi, D. Vikraman, Y.T. Jeong, P. Arunkumar, D.B. Velusamy, T. Maiyalagan, A. Alfantazi, H.S. Kim, High performance, 3D-hierarchical

CoS2/CoSe@ C nanohybrid as an efficient electrocatalyst for hydrogen evolution reaction, J. Alloy Compd., 838 (2020) 155537. https://doi.org/10.1016/j.jallcom.2020.155537

[23] L. Sun, H. Xu, Z. Cheng, D. Zheng, Q. Zhou, S. Yang, J. Lin, A heterostructured WS2/WSe2 catalyst by heterojunction engineering towards boosting hydrogen evolution reaction, Chem. Eng. J., 443 (2022) 136348. https://doi.org/10.1016/j.cej.2022.136348

[24] M. Yang, Z. Jin, C. Wang, X. Cao, X. Wang, H. Ma, H. Pang, L. Tan, G. Yang, Fe Foam-Supported FeS2-MoS2 electrocatalyst for N2 reduction under ambient conditions, ACS Appl. Mater. Interfaces, 13 (46) (2021) 55040-55050. https://doi.org/10.1021/acsami.1c16284

[25] L. Wang, H. Wu, S. Xi, S.T. Chua, F. Wang, S.J. Pennycook, Z.G. Yu, Y. Du, J. Xue, Nitrogen-doped cobalt phosphide for enhanced hydrogen evolution activity, ACS Appl. Mater. Interfaces, 11 (19) (2019) 17359-17367. https://doi.org/10.1021/acsami.9b01235

[26] Y. Ni, X. Ma, S. Wang, Y. Wang, F. Song, M. Cao, C. Hu, Heterostructured nickel/vanadium nitrides composites for efficient electrocatalytic hydrogen evolution in neutral medium, J. Power Sources, 521 (2022) 230934. https://doi.org/10.1016/j.jpowsour.2021.230934

[27] Y. Ling, F. Luo, Q. Zhang, K. Qu, L. Guo, H. Hu, Z. Yang, W. Cai, H. Cheng, Tungsten carbide hollow microspheres with robust and stable electrocatalytic activity toward hydrogen evolution reaction, ACS Omega, 4 (2) (2019) 4185-4191. https://doi.org/10.1021/acsomega.8b03449

[28] Y. Tang, C. Yang, M. Sheng, X. Yin, W. Que, Synergistically Coupling phosphorus-doped molybdenum carbide with Mxene as a highly efficient and stable electrocatalyst for hydrogen evolution reaction, ACS Sust. Chem. Eng., 8 (34) (2020) 12990-12998. https://doi.org/10.1021/acssuschemeng.0c03840

[29] O.A. Fakayode, B.A. Yusuf, C. Zhou, Y. Xu, Q. Ji, J. Xie, H. Ma, Simplistic two-step fabrication of porous carbon-based biomass-derived electrocatalyst for efficient hydrogen evolution reaction, Energy Conv. Manag., 227 (2021) 113628. https://doi.org/10.1016/j.enconman.2020.113628

[30] S. Min, Y. Duan, Y. Li, F. Wang, Biomass-derived self-supported porous carbon membrane embedded with Co nanoparticles as an advanced electrocatalyst for efficient and robust hydrogen evolution reaction, Renew. Energy, 155 (2020) 447-455. https://doi.org/10.1016/j.renene.2020.03.164

[31] S. Sekar, A.T.A. Ahmed, D.H. Sim, S. Lee, Extraordinarily high hydrogen-evolution-reaction activity of corrugated graphene nanosheets derived from biomass rice husks, Int. J. Hydrogen Energy, 47 (95) (2022) 40317-40326. https://doi.org/10.1016/j.ijhydene.2022.02.233

[32] H.H. Fu, L. Chen, H. Gao, X. Yu, J. Hou, G. Wang, F. Yu, H. Li, C. Fan, Y.L. Shi, X. Guo, Walnut shell-derived hierarchical porous carbon with high performances for electrocatalytic hydrogen evolution and symmetry supercapacitors, Int. J. Hydrogen Energy, 45 (1) (2020) 443-451. https://doi.org/10.1016/j.ijhydene.2019.10.159

[33] Y. Fu, J. Pan, G. Xiao, J. Niu, W. Fu, J. Wang, Y. Zheng, C. Li, Difunctional hierarchical CoxP QDs-MoS2@ Ni3S2/NF nanostructure as advanced electrocatalyst for water

electrolysis, J. Mater. Sci. Mater. Electron., 32 (12) (2021) 16126-16138. https://doi.org/10.1007/s10854-021-06161-5

[34] Y. Liu, X. Zhang, W. Zhang, X. Ge, Y. Wang, X. Zou, X. Zhou, W. Zheng, MXene-based quantum dots optimize hydrogen production via spontaneous evolution of Cl-to O-terminated surface groups, Energy Environ. Mater., (2022) e12438. https://doi.org/10.1002/eem2.12438

[35] L. Huo, B. Liu, Z. Gao, J. Zhang, 0D/2D heterojunctions of molybdenum carbide-tungsten carbide quantum dots/N-doped graphene nanosheets as superior and durable electrocatalysts for hydrogen evolution reaction, J. Mater. Chem. A, 5 (35) (2017) 18494-18501. https://doi.org/10.1039/C7TA02864D

[36] J. Jiang, R. Sun, X. Huang, H. Cong, J. Tang, W. Xu, M. Li, Y. Chen, Y. Wang, S. Han, H. Lin, CoS2 quantum dots modified by ZIF-67 and anchored on reduced graphene oxide as an efficient catalyst for hydrogen evolution reaction, Chem. Eng. J., 430 (2022) 132634. https://doi.org/10.1016/j.cej.2021.132634

[37] P. Kumar, A.P. Murthy, L.S. Bezerra, B.K. Martini, G. Maia, J. Madhavan, Carbon supported nickel phosphide as efficient electrocatalyst for hydrogen and oxygen evolution reactions, Int. J. Hydrogen Energy, 46 (1) (2021) 622-632. https://doi.org/10.1016/j.ijhydene.2020.09.263

[38] P. Xu, J. Zhang, Z. Ye, Y. Liu, T. Cen, D. Yuan, Co doped Ni0. 85Se nanoparticles on RGO as efficient electrocatalysts for hydrogen evolution reaction, Appl. Surf. Sci., 494 (2019) 749-755. https://doi.org/10.1016/j.apsusc.2019.07.231

[39] R.V. Digraskar, V.S. Sapner, A.V. Ghule, B.R. Sathe, CZTS/MoS2-rGO Heterostructures: an efficient and highly stable electrocatalyst for enhanced hydrogen generation reactions, J. Electroanal. Chem., 882 (2021) 114983. https://doi.org/10.1016/j.jelechem.2021.114983

[40] Q. Sarmad, U.M. Khan, M.M. Baig, M. Hassan, F.A. Butt, A.H. Khoja, R. Liaquat, Z.S. Khan, M. Anwar, M.A. SA, Praseodymium-doped Sr2TiFeO6-δ double perovskite as a bi-functional electrocatalyst for hydrogen production through water splitting, J. Environ. Chem. Eng., 10 (3) (2022) 107609. https://doi.org/10.1016/j.jece.2022.107609

[41] C. Cui, R. Cheng, H. Zhang, C. Zhang, Y. Ma, C. Shi, B. Fan, H. Wang, X. Wang, Ultrastable MXene@ Pt/SWCNTs' nanocatalysts for hydrogen evolution reaction, Adv. Funct. Mater., 30 (47) (2020) 2000693. https://doi.org/10.1002/adfm.202000693

[42] J. Zhu, L. Cai, X. Yin, Z. Wang, L. Zhang, H. Ma, Y. Ke, Y. Du, S. Xi, A.T. Wee, Y. Chai, Enhanced electrocatalytic hydrogen evolution activity in single-atom Pt-decorated VS2 nanosheets, ACS Nano, 14 (5) (2020) 5600-5608. https://doi.org/10.1021/acsnano.9b10048

[43] B. Ruqia, S. Choi, Catalytic surface specificity on Pt and Pt-Ni (OH)2 electrodes for the hydrogen evolution reaction in alkaline electrolytes and their nano-scaled electro-catalysts, ChemSusChem, 11 (2018) 2643-2653. https://doi.org/10.1002/cssc.201800781

[44] Y. Yao, Engineering the electronic structure of submonolayer Pt on intermetallic Pd3Pb via charge transfer boosts the hydrogen evolution reaction, in Controllable Synthesis and Atomic Scale Regulation of Noble Metal Catalysts (pp. 93-116), Singapore: Springer Singapore, 2022. https://doi.org/10.1007/978-981-19-0205-5_4

[45] W. Cheng, H. Zhang, D. Luan, X.W. Lou, Exposing unsaturated Cu1-O2 sites in nanoscale Cu-MOF for efficient electrocatalytic hydrogen evolution, Sci. Adv., 7 (18) (2021). https://doi.org/10.1126/sciadv.abg2580

[46] J. Hao, W. Yang, Z. Zhang, J. Tang, Metal-organic frameworks derived CoxFe1−xP nanocubes for electrochemical hydrogen evolution, Nanoscale, 7 (25) (2015) 11055-11062. https://doi.org/10.1039/C5NR01955A

[47] H.H. Do, C.C. Nguyen, D.L.T. Nguyen, S.H. Ahn, S.Y. Kim, Q. Van Le, MOF-derived NiSe2 nanoparticles grown on carbon fiber as a binder-free and efficient catalyst for hydrogen evolution reaction, Int. J. Hydrogen Energy, 47 (98) (2022) 41587-41595. https://doi.org/10.1016/j.ijhydene.2022.04.127

[48] G. Zhao, B. Wang, Q. Yan, X. Xia, Mo-doping-assisted electrochemical transformation to generate CoFe LDH as the highly efficient electrocatalyst for overall water splitting, J. Alloys Compd., 902 (2022) 163738. https://doi.org/10.1016/j.jallcom.2022.163738

[49] H. Yang, Z. Chen, P. Guo, B. Fei, R. Wu, B-doping-induced amorphization of LDH for large-current-density hydrogen evolution reaction, Appl. Catal. B Environ., 261 (2020) 118240. https://doi.org/10.1016/j.apcatb.2019.118240

[50] Z.H. Zhang, Z.R. Yu, Y. Zhang, A. Barras, A. Addad, P. Roussel, L.C. Tang, M. Naushad, S. Szunerits, R. Boukherroub, Construction of desert rose flower-shaped NiFe LDH-Ni3S2 heterostructures via seawater corrosion engineering for efficient water-urea splitting and seawater utilization, J. Mater. Chem. A, 11 (36) (2023) 19578-19590. https://doi.org/10.1039/D3TA02770H

Emerging Materials for Next Frontier Energy and Environment Applications   Materials Research Forum LLC
Materials Research Foundations 170 (2024) 60-93   https://doi.org/10.21741/9781644903292-4

Chapter 4

Role of biomass derived functionalized carbon as potential electrode materials for supercapacitor applications

S. Pavithra[1], K. Pramoda[1], A. Sakunthala[2], Rangappa S. Keri[1*]

[1]Centre for Nano and Material Sciences, Jain (Deemed-to-be-University), Jain Global Campus, Kanakapura, Bangalore-562112, Karnataka, India

[2]Solid State Ionics lab, Division of Physical Sciences, Karunya Institute of Technology and Sciences, Coimbatore 641114, Tamil Nadu, India

* sk.rangappa@jainuniversity.ac.in, keriphd@gmail.com

Abstract

Because of their many feedstock options, natural abundance, high temperature stability, more effective resistance to corrosion, simplicity in the process, compatibility with composite materials and hybrid structures, and relative affordability, functionalized carbon materials derived from biomass are becoming an increasingly popular trend as supercapacitor electrode materials. A range of renewable basic feedstocks, such as residual forestry and agricultural resources, industrial byproducts, organic waste from communities, aquatic items, etc., are used as source materials to create biocarbon. The kind of biomass, various synthesis processes, activation procedures, and carbonization strategies all have a significant influence on the the physical and chemical properties, structural, and functional characteristics of the biocarbon materials. These elements directly affect how well the synthetic electrode materials operate electrochemically. This chapter provides an in-depth investigation of the latest developments in the use of functionalized carbon materials produced from biomass for supercapacitor applications, as well as a discussion of the challenges and opportunities that lay beyond.

Keywords

Biomass, Carbon Electrodes, Carbonization, Activation, Supercapacitor

Contents

Role of biomass derived functionalized carbon as potential electrode materials for supercapacitor applications 60

1.   **Introduction**...61

2.   Factors affecting the performance of carbon generated from biomass as an electrode in supercapacitors 62

3.   Strategies to enhance the performance of carbon compounds obtained from biomass in supercapacitors 69

4.   Future Outlook and Conclusion 79

Acknowledgement..79

Reference 79

1.   Introduction

Over the past few decades, there has been a noticeable increase in the world's energy consumption due to factors such as population expansion and increased industrialization. Due to the fact that fossil fuels, a significant source of energy, are quickly running out, attention must be directed on finding alternate energy sources. Researchers are creating a range of energy storage solutions to address the growing demand for energy and the diminishing reserves of fossil resources. As seen in Fig. 1(a), supercapacitors have recently been regarded as one of the most intriguing developments among the numerous energy storage technologies. The most promising technologies are supercapacitors because to their high power density, extended cyclic stability, quick charge-discharge process, increased safety, lack of overcharging risk, etc. Low energy density, high self-discharging rate, high cost of raw materials, and very short duration of power supply are some of the major challenges to be considered while designing a supercapacitor. Based on their energy storage mechanisms (Fig 1(b)), supercapacitors are classified into three types Electrochemical double-layer capacitors (EDLCs), Pseudocapacitors, and Hybrid capacitors. EDLCs store charges based on the charge accumulation mechanism and are highly dependent on interface charge segregation, where no oxidation-reduction reaction takes place and the electrolyte and electrode material are purely electrostatic and non-Faradaic (Fig 1(c)). In contrast to EDLCs, pseudocapacitors use quick and reversible oxidation-reduction events (Faradaic reactions) and ions move into the structure of the electrode material. The materials that primarily contribute to pseudocapacitance are conducting polymers, metal sulfides, metal nitrides, metal hydroxides, and transition metal oxides. When one electrode of an EDLC is combined with other electrodes made of pseudocapacitive materials, a hybrid or asymmetric supercapacitor with higher capacitance is created. This is because one electrode can exhibit electrostatic capacitance, while the other can display electrochemical capacitance.

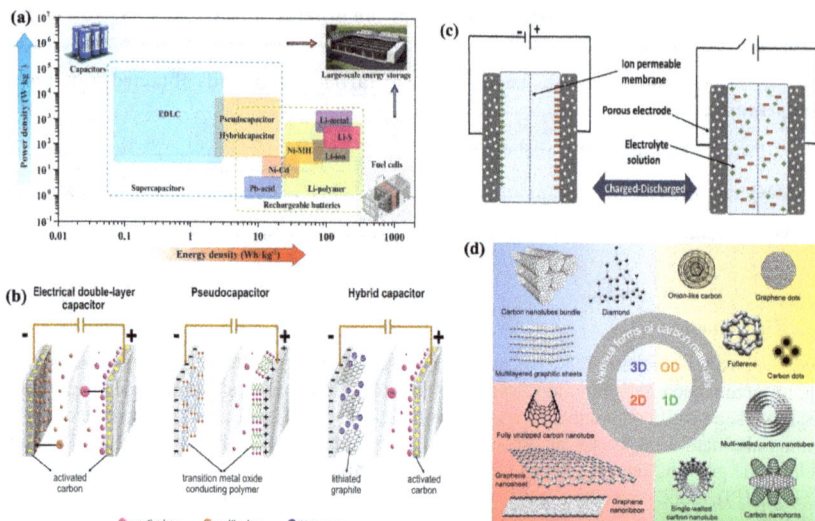

Figure 1 (a) Ragone plot [1] (b) Classification of supercapacitors storage mechanisms [2] (c) Energy storage mechanism in EDLCs [3] (d) Representation of carbon materials in various forms [4]

Various forms of carbon materials like activated carbon, carbon aerogel, graphene, etc based on their dimensions (Fig 1(d)) are used as EDLC materials. Carbon derived from biomass has its potential use in various applications like gas separation [5], pollutants removal [6], electrocatalysts [7], fuel cells [8], solar cells [9], water splitting [10], batteries [11], supercapacitors [12], etc. Some of the important qualities and parameters to be considered while designing a high-performance electrode material for a supercapacitor are specific surface area (SSA), pore structure, electrical conductivity, mechanical strength, wettability of electrolytes on the surface, electrochemical stability [13]. On considering these facts, carbon materials derived from biomass are said to be a suitable material as biomass-derived carbon has the advantages of a larger surface area and good porous structure [14]. Biomass can be classified as agricultural waste, animal waste, forest waste, industrial waste, energy crops, surplus grains, municipal solid waste, sewage, livestock manure, etc which can be effectively used as a source of energy [3, 15]. This chapter includes a thorough discussion of activation procedures, the synthesis approaches, and functional improvement strategies.

2.   Factors affecting the performance of carbon generated from biomass as an electrode in supercapacitors

Synthesis methods of carbon from biomass, nature of biomass, different activation methods and activation agents, and various physiochemical characteristics such as particle size, shape, structure, porosity, functional groups, degree of graphitization, ash content, etc play a major role in influencing the performance of biomass-derived carbon as an electrode for supercapacitor application. Among these porosity is one of the major aspects to be considered while preparing an activated carbon. The straightforward process of creating nano-porous carbon using carbonization and activation techniques is depicted in Fig. 2(a). As seen in Fig. 2(b), the pore structure of activated carbon can be broadly categorized based on their pore size. Since the flow of ions from the electrolyte is a fundamental factor in determining performance of the device, different pore shapes of the activated carbon have a significant impact on the electrochemical performance. The function of various pore shapes and how well they interact with the ions from the electrolyte are shown in Fig. 2(c). It can be seen from part (i) of Fig 2(c) that ultra-micropores are not suitable because of the very low pore size. From parts (ii and iii) of Fig 2(c) it is seen that micro and macropores show good accessibility for electrolyte ions. So it is necessary to design a proper pore structure to maximize accessibility by electrolyte ions. From part (iv) of Fig 2(c) it is seen that the utilization of both macro and micro pores will enhance the accessibility by electrolyte and result in the higher specific capacitance of the synthesized activated carbon. To design and tune the pore structures it is important to choose the proper and wise synthesis methods and activating agents.

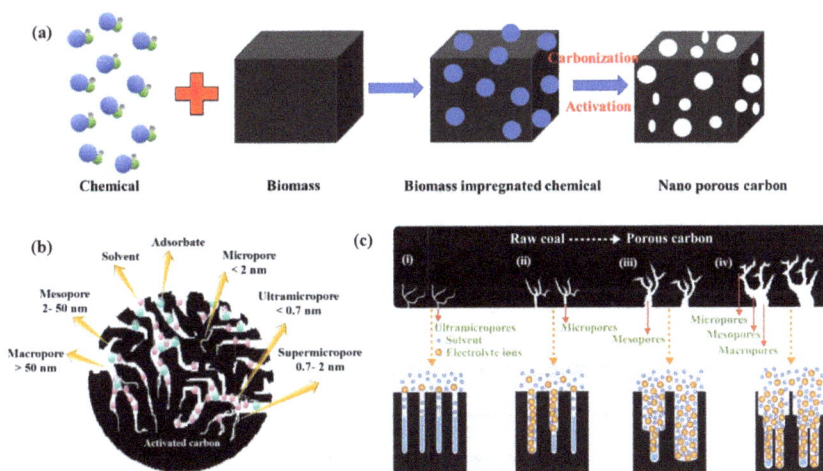

Figure 2 (a) Mechanism of the creation of biomass-derived porous activated carbon [16] (b) Pore structure of activated carbon [13] (c) Influence of pore structure on electrolyte accessibility [1]

The most widely utilized methods for converting biomass into usable carbonaceous material are thermochemical conversion and biomass conversion. Thermochemical conversion methods are the most widely utilized approach for producing carbon materials sourced from biomass for supercapacitor applications, instead of both approaches. Fig 3 (a) illustrates the thermochemical conversion process and the nature of the products obtained. Thermochemical conversion techniques are generally classified into two types dry and wet thermochemical conversion techniques. Dry thermochemical conversion includes processes like pyrolysis, carbonization, liquefication, and torrefaction whereas wet thermochemical conversion includes processes like wet pyrolysis, hydrothermal carbonization, liquefication, and torrefaction. It should be noted that the percentage of yield generally depends on the temperature and pressure of the reaction taking place, the nature of the biomass, biomass loading, and the residence time of the reaction [17, 18].

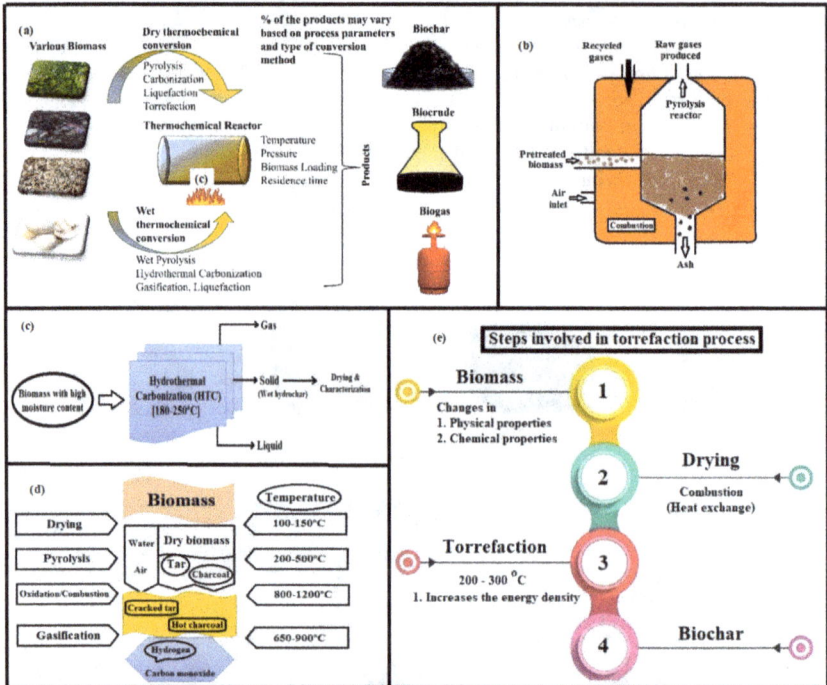

Figure 3 (a) Thermochemical conversion types and products obtained [18] (b) Mechanism of pyrolysis [17] (c) Hydrothermal carbonization procedure [17] (d) Process of gasification [17] (e) Steps involved in torrefaction process [17]

As seen in Fig 3(b), *pyrolysis* is the process of performing thermal degradation in an oxygen-free environment at temperatures between 250-900°C. The type of pyrolysis can be either fast or slow, depending on a number of variables including temperature, pressure, residence time, and heating rate. Dazhi Zhang et al. [19] looked into the effects of the temperature of pyrolysis on the surface area, porousness, ion-transfer kinematics and electrochemical behaviour of carbon materials made from reed residue. The pyrolysis of reed residue at 600°C produced a high surface area of 2074.72

m^2/g and the largest pore volume of 0.930 cm^3/g. These results were followed by a significant specific capacitance of 228 F/g at 1 A/g and 98% capacitance retaining after 8000 cycles at 5 A/g. The study examined different pyrolysis temperatures between 500 and 800 °C. The influence of heating rates at 8, 13, and 30 °C/min was studied on Agave angustifolia leaves-derived carbon materials by Ayala-Cortés et al. [20] and it has been observed that higher surface area values and electrochemical performance were achieved at lower pyrolysis heating rates.

By lowering the oxygen and hydrogen through dehydration and decarboxylation at temperatures between 180-200 °C, a thermochemical process (Fig 3 (c)) creates carbon from biomass known as *hydrothermal carbonisation*. Numerous benefits, including a low use of energy, environmental friendliness, high purity, minimal ash content, controlled size, uniform porosity, and an abundance of surface functional groups, come with this straightforward procedure [1]. It is to be noted that based on the reaction temperatures various materials are obtained and different terms are given to the process as [21]

- Hydrothermal carbonization – below 250°C – biomass derived carbon
- Hydrothermal liquefaction – 250-400°C – bio-oil
- Hydrothermal gasification – above 400°C – CO, CO_2, H_2 and CH_4

The physiochemical characteristics of the produced carbon compounds are determined in large part by a number of parameters, including pH, residence time, temperature, pressure, solid load, and hydrous conditions [22]. On the activated carbon made from salacca peel, Stenny et al. [23] investigated the effects of both temperature (225 °C and 250 °C) as well as pressure (subcritical condition (liquid state) and saturated mixture (vapor-liquid)) at a pressure of 50 bar. When compared to other pressure and temperature conditions, the activated carbon produced during the hydrothermal carbonization process at a temperature of 250 °C in subcritical conditions demonstrated a higher specific surface area of 2907.31 m^2/g due to the existence of significant oxygenated functional groups, leading to in the high specific capacitance.

Gasification is the process of converting biomass into various syngas like CO, CO_2, H_2, and CH_4 at higher temperatures and pressures. It is of two types called air gasification and steam gasification if the air or steam respectively is utilized as a gasification agent. As seen in Fig 3(d), gasification is a four step process that undergoes at different temperature ranges as drying, pyrolysis, oxidation/reduction, and gasification in the presence if gasification agents. The temperature and agents used in gasification, the ratio of gasifying agent to biomass, the moisture content of biomass, reactor design and configuration, the addition of catalysts, and other factors all affect the gasification process [24].

Torrefaction is a simple pyrolysis process that takes place at 200-300°C and has low moisture biomass products. It is a four step process as shown in Fig 3(e). Torrefaction is of three types steam torrefaction - which uses steam at 260 °C for 10 min, wet torrefaction - which uses water at 180-260 °C for 5-240 min and combustion torrefaction- which uses gases.

Emerging Materials for Next Frontier Energy and Environment Applications   Materials Research Forum LLC
Materials Research Foundations 170 (2024) 60-93   https://doi.org/10.21741/9781644903292-4

Table 1 Thermochemical conversion techniques and their process conditions [17]

Technique	Temperature (°C)	Residence time	Yield of biochar (%)
Pyrolysis (slow)	300-700	< 2 s	35
Pyrolysis (fast)	500-1000	Hour-day	12
Hydrothermal carbonization	180-300	1-16 h	50-80
Gasification	750-900	10-20 s	10
Torrefaction	900	10-60 min	80

The process of attempting to improve the surface area and porosity employing various techniques is called activation. There are various activation techniques like physical, chemical, gaseous, physiochemical, microwave-induced activation, and dual activation as shown in Fig 4(a), and the carbon product obtained after activation is called activated carbon (AC). Selecting the appropriate activation agents is crucial since they have a direct impact on many physiochemical parameters such as surface area, degree of graphitization, pore size distribution, bonding state, and surface shape [25].

Physical activation also called thermal activation is a two-step process involving carbonization and activation. During the physical activation, various activating agents like air, steam, and CO_2 are used. Using air as an activating agent is one of the cost effective processes which is generally performed under < 500 °C. Steam is easily available, low cost, and has low activation energy which is generally performed at 800-1200°C. CO_2 activation is the most commonly used method which is generally performed at 800-1000°C where the reaction rate is slow and controlled. When steam and CO_2 are used as activating agents, high oxygen-containing surface functional groups and microporous carbons are produced. Aziz Ahmad et al. [26] synthesized activated carbon from date seeds by subjecting the samples to carbonization at 850 °C for 1 h under the N_2 environment CO_2 physical activation method. The BET surface area was found to be 659.56 m^2/g for the sample which is activated by CO_2 and only 35.21 m^2/g for the sample which is only carbonized and not activated by CO_2. This shows that CO_2 activation enhances the surface area and pore formation and assembled all-solid-state SCs delivered a high specific capacitance of 138.12 F/g at 5 mV/s and showed an energy density of 9.6 Wh/kg with a power density of 87.86 W/kg. L. Chen et al. [27] synthesized activated carbon from rice and also evaluated the influence of steaming with subsequent carbonization processes at 750, 800, 850, and 900 °C (Fig 4b). It can be seen that the steamed rice produced an activated carbon with a large surface area and a rational pore size distribution, because of the honeycomb-like macrostructure caused by the penetration of water molecules into the starch granules forming a network hydrogel.

Chemical activation is the process of activating the carbon precursors using chemical reagents. Selective removal of the non-graphitic part of the carbon occurs by degradation, dehydration, and complexation reactions in chemical activation when compared to physical activation. Various reagents like alkaline materials (KOH, NaOH), acid materials (H_3PO_4, HNO_3, HCl), and various transition metal salts ($CaCl_2$, $ZnCl_2$) are used as chemical activators. This type of activation helps enhance the yield, surface area, and pores by properly tuning the precursor mass ratio, nature of the activating agent, temperature, and time of the activation. The function of several activation agents, such as KOH, $ZnCl_2$, NaOH, H_3PO_4, HCl, and $FeCl_3$, in the production of activated carbon from free-of-cost leftover coffee grounds was investigated by Chiu Y et al [28]. No aggregation and nanoscale carbon were obtained when KOH and NaOH were used. The activated carbon prepared using KOH activator showed a higher surface area of 1250 m^2/g when compared to other reagents which reflected in the high specific capacitance as shown in Fig 4c. Additionally, commercial activated carbon and activated carbon made with a KOH activator were used to fabricate a symmetric supercapacitor. The device constructed using carbon prepared using a KOH activator showed 105.3 F/g with 133% retention after 8000 cycles whereas it was only 21.8 F/g for commercial activated carbon as shown in Fig 4 e, d. The following are the chemically induced reactions that are conducted with KOH [28], and Fig. 4(g) illustrates the chemical reaction mechanism for activation of KOH during biomass pyrolysis.

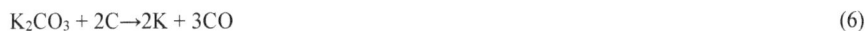

$$4KOH + (-CH_2) \rightarrow K_2CO_3 + K_2O + H_2 \tag{1}$$

$$6KOH + 2C \rightarrow 2K_2CO_3 + 2K + 3H_2 \tag{2}$$

$$K_2CO_3 \rightarrow CO_2 + K_2O \tag{3}$$

$$4KOH + C \rightarrow 4K + CO_2 + 2H_2O \tag{4}$$

$$K_2O + C \rightarrow 2K + CO \tag{5}$$

$$K_2CO_3 + 2C \rightarrow 2K + 3CO \tag{6}$$

Apart from this, the combination of dual activators is also done to enhance the surface area, and pore structure of the activated carbon. Khu Le Van et al. [29] studied the effect of dual alkali hydroxide activators on the pore size, surface area, thermal stability, and electrochemical performance. Various KOH impregnation ratios:NaOH: For the examination of dual alkali hydroxide activators, rice husk char (wt. ratio) such as 4:0:1, 0:3:1, 1:2:1, 1.5:2:1, 2:2:1, 2: 1.5:1, and 2:1:1 were taken into consideration. On equal loading of KOH: NaOH activators (RH-K2N2), a high specific surface area of 2990 m^2/g with the total pore volume of 2990 m^2/g where the micropore and mesopore contribution are 2747 and 243 m^2/g respectively. This low amount of mesopores and surface disorder led to a decrease in the ohmic drop responsible for internal resistance [29]. RH-K2N2 delivered a high specific capacitance of 225 F/g at 0.2 A/g in 0.5 M K_2SO_4 electrolyte.

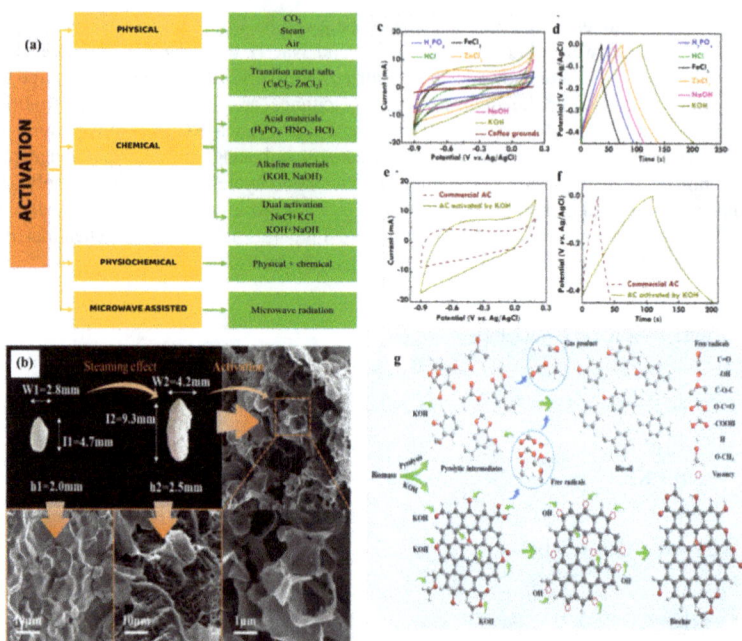

Figure 4 (a) Different types of activation methods (b) Based on the "steaming effect," a hierarchical porous carbon nanosheet fabrication technique was presented using rice [27], Comparison of various activation agents on the three electrode cell based electrochemical performance by (c) CV curves (d) GCD curves, Comparison of commercial activated carbon and KOH activated carbon by two electrode cell based electrochemical performance by (e) CV curves (f) GCD curves [28], (g) Potential chemical route for the activation of KOH during biomass pyrolysis [30]

Another activation technique that combines chemical and physical activation techniques is called *Physiochemical activation*. Chemical activators initiate the formation of porosity followed by physical activation [31]. *Canna indica* flowers were used to produce activated carbon using the physiochemical activation method by D. Navaneethan et al. [32]. Initially, 1 M H_2SO_4 for 24 h was used for chemical activation followed by heating at 300 °C for 2 h in a muffle furnace as a part of physical activation.

Microwave-assisted activation is another versatile technique for the activation of carbon. The internal and volumetric heating strategies speed up chemical processes at lower activation temperatures and, as a result, minimize treatment times and energy usage [31]. Casuarina bark was converted to carbon by T. Gajalakshmi et al. [33] through chemical activation and straightforward microwave pyrolysis. The activation was carried out for three minutes at 1200 W after being mixed with KOH (1:3). With both meso and microporous materials, a high surface area of 4588 m^2/g was attained, and at 0.5 A/g, the specific capacitance was 333 F/g.

Emerging Materials for Next Frontier Energy and Environment Applications   Materials Research Forum LLC
Materials Research Foundations 170 (2024) 60-93   https://doi.org/10.21741/9781644903292-4

3.   Strategies to enhance the performance of carbon compounds obtained from biomass in supercapacitors

The following techniques are being used in supercapacitors to improve their efficiency in relation to high energy density and power density with extended cycle life, carbons generated from biomass [34].

- Functionalisation and Doping
- Surface Hydrophobization
- Hybrid composites with other carbon based materials
- Bifunctional hybrid composites
- Optimization of binders in the electrodes
- Salt and solvent optimization of electrolytes
- Regulation of potential window

Functionalization and doping of biomass-derived carbon materials with various heteroatoms like nitrogen (N), boron (B), oxygen (O), fluorine (F), sulfur (S), and phosphorous (P) effectively enhance the chemical functionalities at the surface. This helps in better conductivity by improving the active sites for reaction and induces the pseudocapacitance nature [35]. Among these various heteroatoms, nitrogen has gained its interest because of the higher contribution of p electrons to the P system which facilitates the enhanced charge carrier transport and a positive synergy with the carbon lattice [34]. Fig 5 (a) shows the four types of functionalities pyrrolic N (N-5), pyridinic N (N-6), quaternary-N (N-Q), and oxidized N (N-X) that exist in biochar [35]. It is reported that N-5 and N-6 are responsible for inducing the pseudocapacitance by creating effects in the electrolytes and increasing more active sites, N-Q enhances the conductivity by facilitating the charge transfer in the electrode-electrolyte interface, N-X supports the wettability [36–39]. Fig 5(a): (i)-(iv) shows the reaction mechanisms of N-5, N-6, N-Q, and N-X [35]. Numerous techniques exist for preparing N-doped carbon, including: (1) hydrothermal treatment in conjunction with pyrolysis; (2) one-step approach; and (3) two-step techniques such as pre-carbonization and activation, pre-activation and the process of carbonization and pyrolysis in succession. (4) As illustrated in Fig. 5(b-g), a two-step carbonization process is followed by microwave activation. Using the electrospinning method, flexible self-supporting N-doped porous biomass carbon nanofiber using squid ink with PAN and ZnO loading was explored for the first time by Yan Gao et al. [40]. Numerous parameters, including the activating temperature (600, 700, and 800°C), the mass ratio of KOH (1, 2, and 3), the mass ratio of squid ink powder to PAN (1, 1.5, and 2), and other parameters, are tuned and their impact on flexible asymmetric supercapacitors is investigated. The optimal configuration for a flexible asymmetric supercapacitor was assembled using a combination of NPBC-1.5@PAN-800–2@ZnO and active carbon cathode (AC) with the PVA-KOH gel electrolyte. Yan Gao et al [40] optimized the mass ratio of squid ink powder and PAN to 1.5, the activation temperature to 800°C, and the mass ratio of KOH to 2. After 5000 cycles, an impressive specific capacitance of 102.41 F/g at 1 A/g with 84.48% capacitance maintenance was attained. The key components that contributed to this enhanced supercapacitor performance were a larger carbon layer spacing, a higher degree of graphitization, and a substantial specific surface area of 636.68 m^2/g.

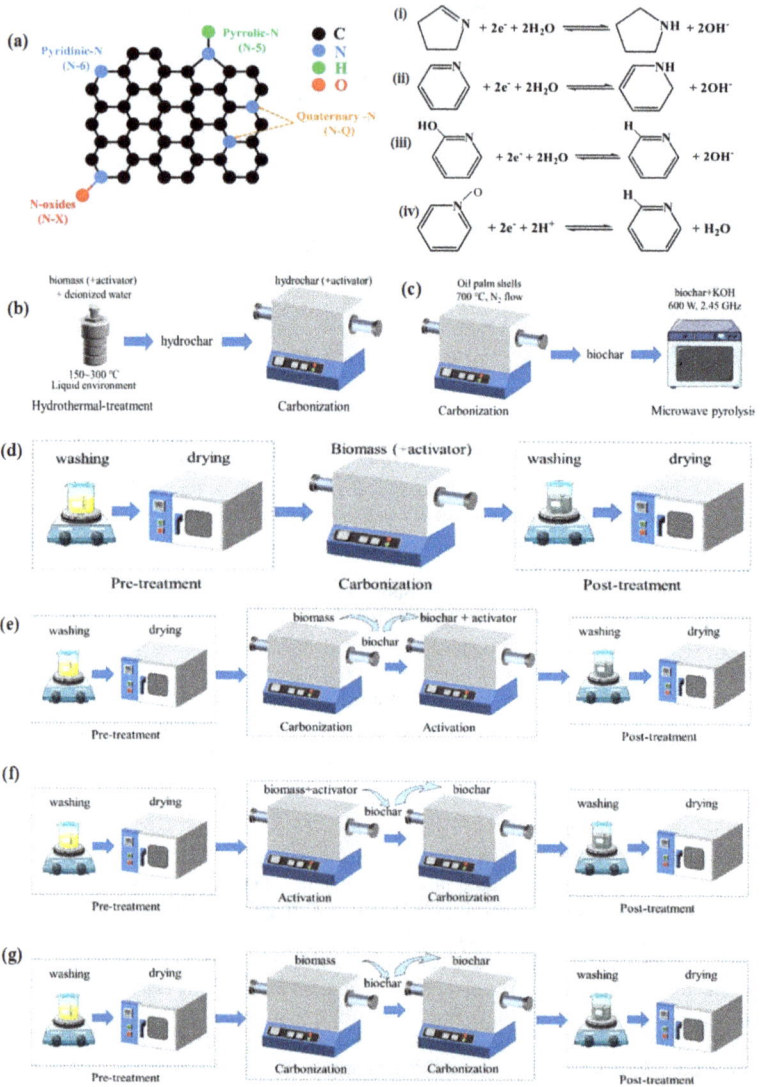

Figure 5 Typical procedure for (a) hydrothermal treatment combined with pyrolysis, (b) two-step procedure of carbonization, followed by microwave activation (c) One-step method, Two-step method (d) pre-carbonization and activation, (e) pre-activation and carbonization, (f) successively pyrolysis [35].

Emerging Materials for Next Frontier Energy and Environment Applications   Materials Research Forum LLC
Materials Research Foundations 170 (2024) 60-93   https://doi.org/10.21741/9781644903292-4

Another important parameter to be considered to enhance the supercapacitor performance is the salt and solvent content of the electrolyte. As the electrolytes play a major role in the charge transfer between electrode and electrolyte interface and the structural stability during cycling it is important to choose a suitable electrolyte [41]. The effect of different aqueous alkaline electrolytes like 6 M KOH and 6 M NaOH on the supercapacitor performance of the activated carbon derived from Phyllanthus emblica leaves was explored by Anant Agrawal et al. [42]. It is seen that the synthesized AC had a larger micropore volume of 0.332 cm^3/g with a mesopore volume of 0.26 cm^3/g. It is important to note that the hydrated ionic radius of K^+ ion is 3.3 Å which is much lower than 3.58 Å of Na^+ and this lower ionic radius is suitable to the adsorption of more K^+ ions. A specific capacitance of 336 F/g at 1 A/g was achieved in KOH. Cycling stability studies were tested for 24,000 cycles in 6 M KOH electrolyte, and EIS were recorded at 1, 8000th, 16,000th, and 24,000th cycle and it is observed that the electrolyte resistance of KOH or internal resistance decreased gradually and the resistance was found to be 1.9, 0.67, 0.7 and 0.66 W respectively as shown in Fig 6(a). This led to the high capacitance retention of 6 M KOH (98.7 %) than 1 M NaOH (97.7 %) and rate capability of 6 M KOH (71.4 %) than 6 M NaOH (49 %) after 8,000 cycles at the 10 A/g. Similarly, Aria Yunita et al. [41] also studied the effect of different electrolytes like 1 M H_2SO_4 (AC-01), 1 M KOH (AC-02), and 1 M NH_3 on the activated carbon from areca nut midrib. Among them, 1 M H_2SO_4 achieved a higher electrochemical performance with 192 F/g at 1 A/g because of the smaller ion size, and higher conductivity of 198 cm^2 W/mol when compared to the other electrolytes. Arisa Phukhrongthung et al. [43] studied the effect of anions (Cl^- and $TFSI^-$) in super-concentrated conditions (water-in-salt) on the oil palm leaf-derived hierarchical porous carbon. When LiTFSI is used as an electrolyte, the voltage window was achieved up to 2.6 V whereas it was only 1.9 V for LiCl. But on considering the specific capacitance value, LiCl showed a higher capacitance of 331 F/g whereas it was only 176 F/g for LiTFSI. It should be noted that the hydrated ion radius of Li^+ is 0.38 nm, Cl^- is 0.32 nm, and $TFSI^-$ is 0.79 nm and it is seen that Cl^- has a lower size than $TFSI^-$. It is seen that the diffusion of Li^+ and Cl^- is higher into the small channels leading to the higher capacitance value as seen in Fig 6(b). Also, Fig 6(c) shows that the high access to the inner surface (pores and grain boundaries) was achieved by the LiCl due to its smaller anion size.

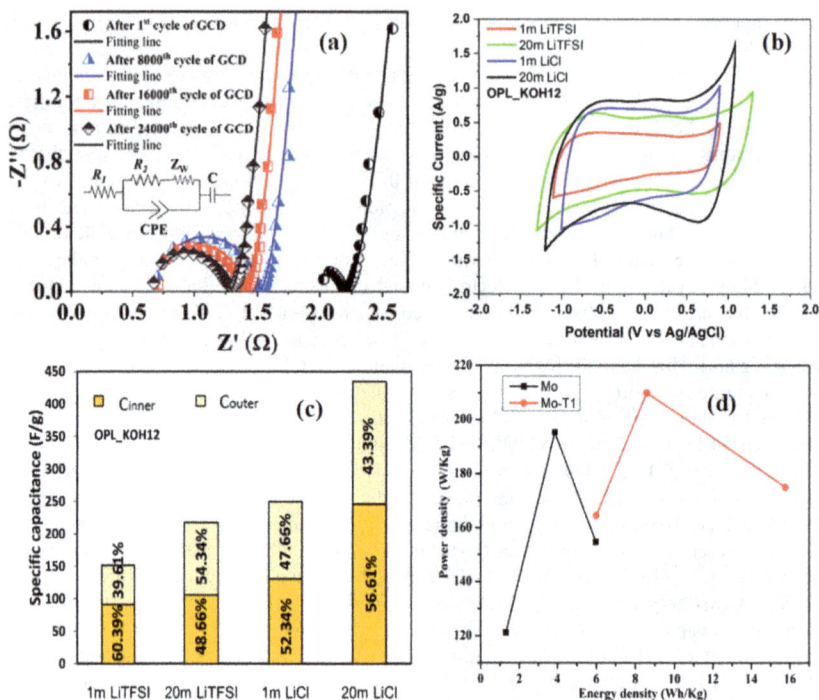

Figure 6 (a) Nyquist curves [42] (b) Comparison of CV curves in different electrolytes [43] (c) The particular capacitance, which uses the Trasatti method to provide the outer capacitance obtained at an extremely fast scan rate and the overall capacitance collected at an extremely low scan rate. [43] (d) Ragone plots illustrating Mo and Mo-T1 performance as possible anode choices for supercapacitors [44]

Hybrid composites with other carbon-based materials, the mixture of two biomass-based materials, and metal oxides are other efficient ways to enhance the supercapacitor performance. Husks of Moringa bark (Mo) and its spent tea waste composites (Stw) in different ratios of 100:0, 90:10, 80:20, 70:30, and 50:50 were taken to produce activated carbon by Zeeshan Mujtaba et al. [44]. Mo: Stw composite prepared in a 90:10 ratio (Mo-T1) showed a higher supercapacitor performance of 231.4 F/g at 0.5 A/g which was almost three times higher than the performance of bare Moringa bark husk (Mo). Fig 6(d) shows the comparison of energy and power density of Mo and Mo-T1. A good synergetic effect between Mo and Stw, and an increase in pore volume with 10% Stw addition helped in the superior supercapacitor performance of the composites. But when the addition of Stw increased to 20, 30, and 40 the performance decreased because of the reduction of pore size due to higher graphitization. So it is important to optimize the appropriate percentage of composite to achieve higher supercapacitor performance. Bing Wang et al. [45] used rose branch biomass was used to synthesize hierarchically porous carbon-supported MnOₓ (MnO and MnO₂) electrodes as shown in Fig 7(a). In this work, the supercapacitor performance was improved by altering the operating voltage, fast ion/electron transport, reducing carbon pore blocking, and using MnO_x as it enhances the pseudocapacitance contribution and acts as a buffer to structural changes. The synthesized MnO_2/AC and MnO/AC showed a specific surface area of 769.7 and 1693.6 m^2/g, respectively. The electrochemical performance was recorded using 1 M Na_2SO_4 and Fig 7b shows the energy storage mechanisms associated with the micropore and mesopore of MnOx/AC electrode as physisorption/desorption, chemisorption/ desorption, redox reactions, and intercalation/deintercalation processes. From Fig 7(c) it can be seen that MnO/AC shows higher performance than MnO_2/AC due to the valence states of Mn in the composites and proves that MnO has more potential than MnO_2. Also, the MnO/AC showed 1.8 V of voltage window and 328 F/g at 1 A/g whereas MnO_2/AC showed only 1.6 V and exhibited 235 F/g.

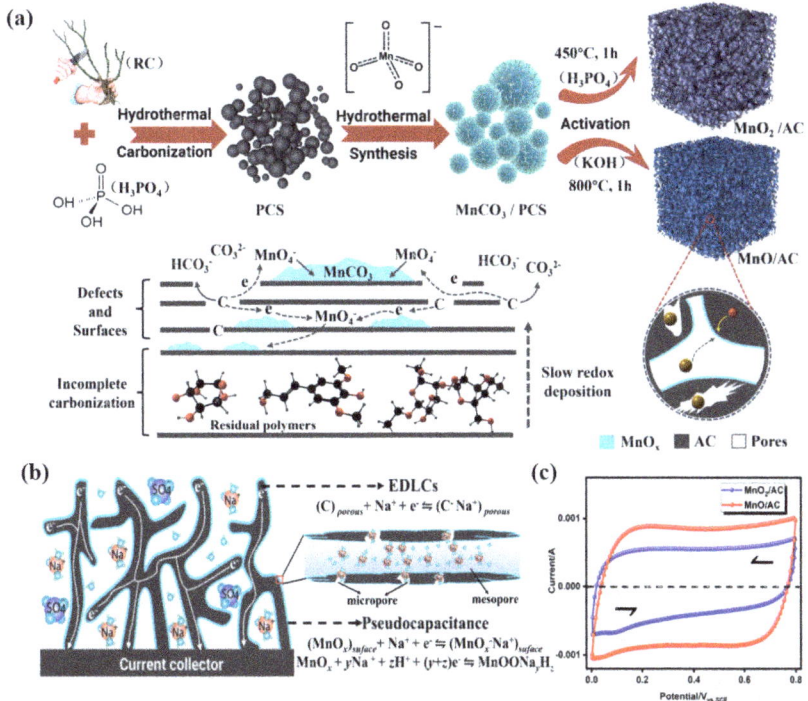

Figure 7 (a) Illustration describing the steps involved in producing MnO2/AC and MnO/AC (b) Energy storage processes connected to the MnOx/AC electrode (c) CV profiles of MnO2/AC and MnO/AC recorded at 2 mV/s [45]

Table 2 Comparison of various biomass derived carbon as electrode materials for supercapacitor applications

Type of carbon material	Biomass precursors	Surface area (m²/g)	Electrolyte	Specific capacitance (F/g) @ current density (A/g)	Retention % @ cycles	Ref
Carbon quantum dots	Spent tea leaves	-	1M H_2SO_4	302 @ 0.5	93.8 (5000)	[46]
Carbon quantum dots	Waste soybean oil	69.50	0.5 M $NaPF_6$	386.5 @ 0.5	-	[47]
Porous carbon	*Phyllanthus emblica*	1244	6 M KOH	336 @ 1	98.7 (8000)	[42]
Activated carbon	Sesamum indicum oil	1158	0.5 M $NaPF_6$	94 @ 0.5 mA/g	82 (2000)	[48]
Activated carbon	Areca nut midrib	-	1M H_2SO_4	192 @ 1	-	[41]
Porous carbon	Oil palm leaf	1840	20 m LiCl	331 @ 10 mV/s	-	[43]
Carbon aerogel	Sugarcane juice	2028	1 M Na_2SO_4	136.86 @ 5 mV/s	70 (5500)	[49]
Carbon	Hyphaene fruit shell	197.99	3 M KOH	337 @ 0.7	100 (3000)	[50]
Activated carbon	Date seeds	-	1M H_2SO_4	160 @ 0.2	95.2 (5500)	[51]
Porous carbon	Rose branches	1693	1 M Na_2SO_4	328 @ 1	96.6 (10,000)	[45]
Carbon	Wood blocks (basswood)	1115.5	5 M H_2SO_4	394	100 (50,000)	[52]
CNT	Wood blocks (basswood)	37.8	6 M KOH	118.5	94.6 (10,000)	[53]
Activated carbon	*Bauhinia variegata* L	-	6 M KOH	165 @ 0.5	86 (5000)	[54]
Porous carbon	*Amygdalus davidiana shells*	1734.58	6 M KOH	208 @ 1	100 (20,000)	[55]
Activated carbon	Triphala seed stones	1233.3	1M H_2SO_4	208.7 @ 1	-	[56]
Porous carbon	Tea seed meal	1764	6 M KOH	384 @ 0.5	-	[57]
Porous carbon	Bamboo cells	2420.6	6 M KOH	403.2 @ 0.5	96 (5000)	[58]
Porous carbon	Bamboo stem cells	1987.76	6 M KOH	508 @ 0.5	-	[59]
Porous carbon	Rice husk	643.48	6 M KOH	78.70 @ 0.5	97.2 (10000)	[60]
Porous carbon	Caragana straw	1779	6 M KOH	427 @ 0.5	97.4 (10000)	[61]
Porous carbon	Wood fibers	593.52	6 M KOH	270 @ 0.5	98.4 (10000)	[62]

Activated carbon	Moringa oleifera and its spent tea waste composites	-	1 M KOH	231.4 @ 0.5	93 (1000)	[44]
Porous carbon	Coconut shell	2143.6	6 M KOH	317 @ 0.5	99.7 (10000)	[63]
Porous carbon	Jujube shells	-	6 M KOH	535 @ 1	-	[64]
Porous carbon	Wheat straw	1961.1	1M H_2SO_4	424.9 @ 1	93.1 (10000)	[65]
Porous carbon	Red dates	1115.7	6 M KOH	341.9 @ 0.5	-	[66]
Activated carbon	Chinese fir wood scraps	420	1M H_2SO_4	14.31F/cm^2 @ 5 mA/cm^2	-	[67]
Porous carbon	*Spirulina platensis*	2923.7	6 M KOH	348 @ 1	94.14 (10000)	[68]
Activated carbon	*Chlorella protothecoides*	-	6 M KOH	233.15 @ 1	-	[69]
Activated carbon	*Dunaliella salina microalgae*	373.40	6 M KOH	284.86 @ 1	92.57 (100)	[70]
Nano carbon	Rice straw	1131.7	6 M KOH	331 @ 0.5	97.6 (10000)	[71]
Activated carbon	Bamboo parenchyma cells	3973	6 M KOH	276 @ 0.1	89.48 (15000)	[72]
Porous carbon	Ganoderma lucidem	-	3 M KOH	242.6 @ 2	84.5 (4000)	[73]
Porous carbon	Sugarcane bagasse	2803.2	6 M KOH	356.4 @ 1	95 (10000)	[74]
Carbon	Balsa wood	548.93	6 M KOH	252.1 @ 0.5	100 (10000)	[75]
Graphene	Kitchen waste	2097	2 M KOH	237.4 @ 0.5	-	[76]
Porous carbon	Garlic peels	3272	6 M KOH	227 @ 10	96 (5000)	[77]
Activated carbon	Eucommia ulmoides wood tar	3354.8	1 M KOH	510 @ 0.2	-	[78]
Porous carbon	Passion fruit husk	1996.3	1 M trifluoroacetic acid	297.1 @ 1	-	[79]
Porous carbon	Semi-coking wastewater	673	6 M KOH	129.5 @ 1	99.83 (10000)	[80]
Porous carbon	Pomelo peel	822.8	6 M KOH	400 @ 0.5	-	[81]
Porous carbon	Potato starch	2267	0.5 M H_2SO_4	727 @ 0.5	92.1 (50000)	[82]
Porous carbon	Basswood blocks	516.44	1M H_2SO_4	203 @ 1mA/ cm^2	96 (10000)	[83]
Porous carbon	Wintersweet-fruit-shell	1489.31	6 M KOH	433 @ 0.5	95.20 (5000)	[84]
Porous carbon	*Linum usitatissimum L.* Root	839	6 M KOH	421 @ 1	-	[85]

Porous carbon	*Linum usitatissimum L.* Stem	851	6 M KOH	434 @ 1	-	[86]
Porous carbon	*Houttuynia cordata*	2491.5	6 M KOH	220.5 @ 0.5	-	[87]
Porous carbon	Palm empty fruit bunches	1496	1M H_2SO_4	121 @ 0.1	-	[88]
Porous carbon	Cottonseed meal	1573	6 M KOH	477 @ 0.1	-	[89]
Activated carbon	Pineapple leaf fibre	643.40	1M H_2SO_4	400 @ 0.1	86 (10000)	[90]
Porous carbon	Bamboo powder	1296	6 M KOH	394 @ 1	-	[91]
Porous carbon	Rotten grapes	2756	1M KOH	375 @ 1	-	[92]
Activated carbon	Ricinus communis shell	1917	3 M KOH	137 @ 1	97.2 (5000)	[93]
Porous carbon	*Annona reticulate*	262.95	1M KOH	573 @ 2 mA/cm^2	92.10 (3000)	[94]
Porous carbon	Oil pressed residue of Pongamia pinnata	811	1 M Na_2SO_4	205 @ 1	-	[95]
Porous carbon	Mixed Egg and Rice Waste	1572.1	6 M KOH	446 @ 1	82.26 (10000)	[96]
Activated carbon	Cilantro plants	2200	6 M KOH	162.4 @ 1	-	[97]
Porous carbon	Litchi seeds	726	1M H_2SO_4	995 @ 1	-	[98]
Porous carbon	Soybean straw	1756.74	1M KOH	220 @ 0.5	98.5 (10000)	[99]
Porous carbon	Lavender straw	1251	6 M KOH	401 @ 1	100 (10000)	[100]
Activated carbon	Rice straw	694.05	3 M KOH	112.05 @ 1	-	[101]
Activated carbon	Date seeds	-	1M KOH	487.5 @ 1	-	[102]
Activated carbon	Date seeds	1045	1M H_2SO_4	298.5 @ 0.5	-	[103]
Porous carbon	Anacardium occidentale Nut-Skin	615	1M H_2SO_4	193 @ 0.5	97 (10000)	[104]
Porous carbon	Waste chips of moso bamboo	961	6 M KOH	256 @ 1	87.45 (5000)	[105]
Activated carbon	Durian shell	520.65	1M KOH	325.2 @ 1	94.79 (10000)	[106]
Porous carbon	Rice husk bio-oil	1905.3	6 M KOH	233.4 @ 0.5	92.3 (10000)	[107]
Activated carbon	Waste hazelnut shells	2092	6 M KOH	149 @ 1	-	[108]
Activated carbon	Palm petiole	1365	1M KOH	309 @ 1	94 (10000)	[109]
Carbon nanosheets	Acorus calamus	3551.07	6 M KOH	119.6 @ 1	100 (1000)	[110]

Activated carbon	Lycopodium clavatum spores	1186	6 M KOH	182.9 @ 0.1	87 (10000)	[111]
Activated carbon	Willow catkin	-	3 M KOH	105 @ 1	98.1 (5000)	[112]
Activated carbon	Willow catkin	-	3 M KOH	173.3 @ 0.7	89.23 (1000)	[113]
Graphitic carbon quantum dot	Lemon juice	-	6 M KOH	284 @ 1	94.5 (10000)	[114]
Porous carbon	Enhydra fluctuant leaves	1082.8	0.5 M H_2SO_4	428 @ 1	-	[115]
Porous carbon	Plane tree fruit fluffs	1582	6 M KOH	208.4 @ 0.2	-	[116]
Activated carbon	Cashew nutshell	898.3	1 M Na_2SO_4	214 @ 1	98 (1000)	[117]
Porous carbon	Sesame capsule shells	12.2183	2 M KOH	222.7 @ 1	92.9 (20000)	[118]
Activated carbon	Orange peels	596	0.5 M H_2SO_4	47.68 @ 0.4	-	[119]
Porous carbon	Waste cotton	1727.9	6 M KOH	273.7 @ 1	97.49 (8000)	[120]
Porous carbon	Salviae miltiorrhizae radix et rhizoma	3410	6 M KOH	291.6 @ 1	-	[121]
N doped Activated carbon	Eucalyptus leaves	1027	2 M KOH	258 @ 0.25	-	[122]
Activated carbon	Pollen typhae	1388.91	1 M H_2SO_4	437 @ 1	90.9 (100,000)	[123]
Activated carbon	Corncobs	212.554	2 M H_2SO_4	132 @ 1	57.23 (1000)	[124]
Porous carbon	Popcorn	65.79	6 M KOH	425 @ 1	92 (10000)	[125]
Hard carbon	Olive leaves	-	6 M KOH	169.6 @ 0.5	96.7 (20000)	[126]
Porous carbon	Palm flower	3362.51	6 M KOH	391.05 @ 1	93.43 (10000)	[127]
Activated carbon	Beta vulgaris L	2200	1 M Na_2SO_4	492 @ 1	-	[128]
Nitrogen-doped carbon	Black liquor and its mixture with wood charcoal	2481	1 M Na_2SO_4	142.23 @ 0.2	99 (1000)	[129]
Carbon aerogels	Liquefied wood	1996.11	6 M KOH	201 @ 0.5	94.11(5000)	[130]
N, O co-doped hierarchical porous carbon	Soybean protein	1350	6 M KOH	248 @ 0.5	-	[131]
N, O co-doped	Waste eucalyptus bark	1719.15	1 M H_2SO_4	483.5 @ 0.5	-	[132]

hierarchical porous carbon						
Porous carbon	Peanut shell waste	550.38	6 M KOH	575.7 @ 0.5	106.5 (10000)	[133]
Activated carbon	Corncob	725	6 M KOH	70 @ 5	94 (1000)	[134]
Activated carbon	Tangerine peel	902	1 M KOH	233 @ 1	-	[135]
Oxygen-rich functionalized porous carbon	Prosopis juliflora	1210.4	1 M Na_2SO_4	161 @ 0.3	-	[136]
Porous carbon	Seeds of Palmyra Palm Tree	-	0.5 M H_2SO_4	276.5 @ 1	63.1 (500)	[137]
Porous carbon	Poplar wood chips	1101.9	1 M KOH	189.3 @ 1	-	[138]
Carbon	Cornstalk	812.84	6 M KOH	133.32 @ 1	-	[139]
Activated carbon	American ginseng	2187	6 M KOH	268 @ 0.1	97.47 (10000)	[140]
Porous carbon	Equisetum ramosissimum Desf weeds	1805.5	6 M KOH	427 @ 0.2	97.8 (10000)	[141]
N/O co-doped hierarchical porous carbon	Orange peels	3253	6 M KOH	352 @ 0.5	99.5 (10000)	[142]
Activated carbon	Eucommia ulmoides Oliver wood	2033.87	1 M H_2SO_4	252 @ 0.2	92.3 (10000)	[143]
Oxygen-enriched hierarchical porous carbons	Lignite (low-rank coal)	1638	6 M KOH	283 @ 0.5	-	[144]
Activated carbon	Red pepper industrial waste	745	6 M KOH	131 @ 0.5	98 (10000)	[145]
N doped porous carbon	Bio-oil distillation residue	1853.53	6 M KOH	442 @ 1		[146]
N/O self-doped hierarchical porous carbons	Shaerhu subbituminous coal	3586	6 M KOH	373 @ 0.5	-	[147]
Porous carbon	Vulcanized natural rubber latex foams	1854.3	1 M H_2SO_4	430 @ 1	-	[148]
Porous carbon	Disposal of Chinese medicine residues	402.42	6 M KOH	221 @ 0.5	100 (10000)	[149]
Porous carbon	Byproduct tar from biomass gasification process	2131.05	6 M KOH	310.4 @ 0.2	91 (5000)	[150]
Porous carbon	Waste antibiotic fermentation residues	948	6 M KOH	310 @ 0.5	95 (10000)	[151]

Porous carbon	Furfural residue	1850	6 M KOH	222.7 @ 0.5	96.43 (5000)	[152]
Porous carbon	Sludge-derived humic acid	1928.09	-	131.2 @ 0.5	92 (1000)	[153]
Porous carbon	Abandoned shuttlecock feathers	3474	6 M KOH	709 @ 0.5	-	[154]
Porous carbon	Heavy-fraction bio-oil	1645	6 M KOH	336 @ 0.5	-	[155]
Porous carbon	Heavy bio-oil	1508	6 M KOH	344 @ 0.5	-	[156]
Activated carbon	Coal liquefaction residual asphaltene	1730	6 M KOH	247 @ 1	-	[157]
Porous carbon	Coke solid waste	2738	6 M KOH	360 @ 1	95.7 (10000)	[158]
Activated carbon	Waste denim	1058	1M H_2SO_4	95.93 @ 1	83.01 (3000)	[159]
Porous carbon	Coal liquefaction residue	1368	6 M KOH	268.5 @ 1	-	[160]
Porous carbon	Coal liquefaction residue	2691	6 M KOH	457 @ 0.5	-	[161]
Porous carbon	Mussel cooking wastewater	1526	Liquid DES	657 @ 1	100 ± 2 (1000)	[162]
Porous carbon	Squid chitin	149.4	Liquid DES	20 @ 1	93.3 (5000)	[163]

4. Future Outlook and Conclusion

In this chapter, the role of biomass in supercapacitor application, different synthesis strategies, different activation types and agents, the influence of physiochemical properties, and strategies to enhance the performance were discussed. Apart from the existing research on biomass-derived activated carbon, there are several research gaps and challenges to be focussed on the upcoming days to make it an efficient material for scale-up and commercialization. Efforts should be taken to check the availability of biomass, develop sustainable and scalable synthesis methods, analysis of economic feasibility and commercial viability, test for real-time applications, tailor the intrinsic properties, and applications of machine learning are to be considered in efficiently utilizing the biomass. In conclusion, the use of biomass as a carbon source is a simple, economical, and efficient approach to developing carbon-based electrodes for supercapacitor applications.

Acknowledgement

Authors acknowledges Jain (Deemed-to-University), Bangalore for the financial support. The author S. Pavithra thanks the Jain University for awarding Post-doctoral fellowship (Ref. No: JU/APP/CRTA/2023/808)

Reference

[1]Dong, D.; Xiao, Y. Recent Progress and Challenges in Coal-Derived Porous Carbon for Supercapacitor Applications. *Chemical Engineering Journal*. Elsevier B.V. August 15, 2023. https://doi.org/10.1016/j.cej.2023.144441.

[2] da Silva, E. P.; Fragal, V. H.; Fragal, E. H.; Sequinel, T.; Gorup, L. F.; Silva, R.; Muniz, E. C. Sustainable Energy and Waste Management: How to Transform Plastic Waste into Carbon Nanostructures for Electrochemical Supercapacitors. *Waste Management*, 2023, *171*, 71–85. https://doi.org/10.1016/j.wasman.2023.08.028.

[3] Rawat, S.; Wang, C. T.; Lay, C. H.; Hotha, S.; Bhaskar, T. Sustainable Biochar for Advanced Electrochemical/Energy Storage Applications. *Journal of Energy Storage*. Elsevier Ltd July 1, 2023. https://doi.org/10.1016/j.est.2023.107115.

[4] Mudassir, M. A.; Kousar, S.; Ehsan, M.; Usama, M.; Sattar, U.; Aleem, M.; Naheed, I.; Saeed, O. Bin; Ahmad, M.; Akbar, H. F.; et al. Emulsion-Derived Porous Carbon-Based Materials for Energy and Environmental Applications. *Renewable and Sustainable Energy Reviews*. Elsevier Ltd October 1, 2023. https://doi.org/10.1016/j.rser.2023.113594.

[5] Hakami, O. Urea-Doped Hierarchical Porous Carbons Derived from Sucrose Precursor for Highly Efficient CO2 Adsorption and Separation. *Surfaces and Interfaces*, 2023, *37*, 102668. https://doi.org/https://doi.org/10.1016/j.surfin.2023.102668.

[6] Chen, Y.; Yang, Y.; Ren, N.; Duan, X. Single-Atom Catalysts Derived from Biomass: Low-Cost and High-Performance Persulfate Activators for Water Decontamination. *Curr Opin Chem Eng*, 2023, *41*, 100942. https://doi.org/https://doi.org/10.1016/j.coche.2023.100942.

[7] Cao, Y.; Sun, Y.; Zheng, R.; Wang, Q.; Li, X.; Wei, H.; Wang, L.; Li, Z.; Wang, F.; Han, N. Biomass-Derived Carbon Material as Efficient Electrocatalysts for the Oxygen Reduction Reaction. *Biomass Bioenergy*, 2023, *168*, 106676. https://doi.org/https://doi.org/10.1016/j.biombioe.2022.106676.

[8] Chu, C.; Liu, J.; Wei, L.; Feng, J.; Li, H.; Shen, J. Iron Carbide and Iron Phosphide Embedded N-Doped Porous Carbon Derived from Biomass as Oxygen Reduction Reaction Catalyst for Microbial Fuel Cell. *Int J Hydrogen Energy*, 2023, *48* (11), 4492–4502. https://doi.org/https://doi.org/10.1016/j.ijhydene.2022.10.262.

[9] Wang, X.; Xu, L.; Ge, S.; Foong, S. Y.; Liew, R. K.; Fong Chong, W. W.; Verma, M.; Naushad, Mu.; Park, Y.-K.; Lam, S. S.; et al. Biomass-Based Carbon Quantum Dots for Polycrystalline Silicon Solar Cells with Enhanced Photovoltaic Performance. *Energy*, 2023, *274*, 127354. https://doi.org/https://doi.org/10.1016/j.energy.2023.127354.

[10] Zhang, W.; Liu, R.; Fan, Z.; Wen, H.; Chen, Y.; Lin, R.; Zhu, Y.; Yang, X.; Chen, Z. Synergistic Copper Nanoparticles and Adjacent Single Atoms on Biomass-Derived N-Doped Carbon toward Overall Water Splitting. *Inorg Chem Front*, 2023, *10* (2), 443–453. https://doi.org/10.1039/D2QI02285K.

[11] Wu, S.; Jin, Y.; Wang, D.; Xu, Z.; Li, L.; Zou, X.; Zhang, M.; Wang, Z.; Yang, H. Fe2O3/Carbon Derived from Peanut Shell Hybrid as an Advanced Anode for High Performance Lithium Ion Batteries. *J Energy Storage*, 2023, *68*, 107731. https://doi.org/https://doi.org/10.1016/j.est.2023.107731.

[12] Xi, Y.; Xiao, Z.; Lv, H.; Sun, H.; Wang, X.; Zhao, Z.; Zhai, S.; An, Q. Template-Assisted Synthesis of Porous Carbon Derived from Biomass for Enhanced Supercapacitor Performance. *Diam Relat Mater*, 2022, *128*, 109219. https://doi.org/https://doi.org/10.1016/j.diamond.2022.109219.

[13] Priya, D. S.; Kennedy, L. J.; Anand, G. T. Emerging Trends in Biomass-Derived Porous Carbon Materials for Energy Storage Application: A Critical Review. *Materials Today Sustainability*. Elsevier Ltd March 1, 2023. https://doi.org/10.1016/j.mtsust.2023.100320.

[14]   Sriram, G.; Kurkuri, M.; Oh, T. H. Recent Trends in Highly Porous Structured Carbon Electrodes for Supercapacitor Applications: A Review. *Energies*. MDPI June 1, 2023. https://doi.org/10.3390/en16124641.

[15]   He, H.; Zhang, R.; Zhang, P.; Wang, P.; Chen, N.; Qian, B.; Zhang, L.; Yu, J.; Dai, B. Functional Carbon from Nature: Biomass-Derived Carbon Materials and the Recent Progress of Their Applications. *Advanced Science*. John Wiley and Sons Inc June 2, 2023. https://doi.org/10.1002/advs.202205557.

[16]   Sahu, R. K.; Gangil, S.; Bhargav, V. K.; Sahu, P.; Ghritalahre, B. Synthesizing Biomass into Nano Carbon for Use in High-Performance Supercapacitors - A Brief Critical Review. *Journal of Energy Storage*. Elsevier Ltd November 20, 2023. https://doi.org/10.1016/j.est.2023.108348.

[17]   Yaashikaa, P. R.; Kumar, P. S.; Varjani, S.; Saravanan, A. A Critical Review on the Biochar Production Techniques, Characterization, Stability and Applications for Circular Bioeconomy. *Biotechnology Reports*, 2020, *28*, e00570. https://doi.org/https://doi.org/10.1016/j.btre.2020.e00570.

[18]   Venkatachalam, C. D.; Sekar, S.; Sengottian, M.; Ravichandran, S. R.; Bhuvaneshwaran, P. A Critical Review of the Production, Activation, and Morphological Characteristic Study on Functionalized Biochar. *Journal of Energy Storage*. Elsevier Ltd September 1, 2023. https://doi.org/10.1016/j.est.2023.107525.

[19]   Zhang, D.; Zhang, Y.; Liu, H.; Xu, Y.; Wu, J.; Li, P. Effect of Pyrolysis Temperature on Carbon Materials Derived from Reed Residue Waste Biomass for Use in Supercapacitor Electrodes. *Journal of Physics and Chemistry of Solids*, 2023, *178*. https://doi.org/10.1016/j.jpcs.2023.111318.

[20]   Ayala-Cortés, A.; Arancibia-Bulnes, C. A.; Villafán-Vidales, H. I.; Lobato-Peralta, D. R.; Martínez-Casillas, D. C.; Cuentas-Gallegos, A. K. Solar Pyrolysis of Agave and Tomato Pruning Wastes: Insights of the Effect of Pyrolysis Operation Parameters on the Physicochemical Properties of Biochar. In *AIP Conference Proceedings*; American Institute of Physics Inc., 2019; Vol. 2126. https://doi.org/10.1063/1.5117681.

[21]   Safaei Khorram, M.; Zhang, Q.; Lin, D.; Zheng, Y.; Fang, H.; Yu, Y. Biochar: A Review of Its Impact on Pesticide Behavior in Soil Environments and Its Potential Applications. *Journal of Environmental Sciences*, 2016, *44*, 269–279. https://doi.org/https://doi.org/10.1016/j.jes.2015.12.027.

[22]   Funke, A.; Ziegler, F. Hydrothermal Carbonization of Biomass: A Summary and Discussion of Chemical Mechanisms for Process Engineering. *Biofuels, Bioproducts and Biorefining*. March 2010, pp 160–177. https://doi.org/10.1002/bbb.198.

[23]   Stenny Winata, A.; Devianto, H.; Frida Susanti, R. Synthesis of Activated Carbon from Salacca Peel with Hydrothermal Carbonization for Supercapacitor Application. *Mater Today Proc*, 2021, *44*, 3268–3272. https://doi.org/https://doi.org/10.1016/j.matpr.2020.11.515.

[24]   Perera, S. M. H. D.; Wickramasinghe, C.; Samarasiri, B. K. T.; Narayana, M. Modeling of Thermochemical Conversion of Waste Biomass – a Comprehensive Review. *Biofuel Research Journal*, 2021, *8* (4), 1481–1528. https://doi.org/10.18331/BRJ2021.8.4.3.

[25]   Doğan, H.; Taş, M.; Meşeli, T.; Elden, G.; Genc, G. Review on the Applications of Biomass-Derived Carbon Materials in Vanadium Redox Flow Batteries. *ACS Omega*. American Chemical Society September 26, 2023, pp 34310–34327. https://doi.org/10.1021/acsomega.3c03648.

[26]   Ahmad, A.; Gondal, M. A.; Hassan, M.; Iqbal, R.; Ullah, S.; Alzahrani, A. S.; Memon, W. A.; Mabood, F.; Melhi, S. Preparation and Characterization of Physically Activated Carbon and Its Energetic Application for All-Solid-State Supercapacitors: A Case Study. *ACS Omega*, 2023, *8* (24), 21653–21663. https://doi.org/10.1021/acsomega.3c01065.

[27]   Chen, L.; Xiang, L. Y.; Hu, B.; Zhang, H. Q.; He, G. J.; Yin, X. C.; Cao, X. W. Hierarchical Porous Carbons with Honeycomb-like Macrostructure Derived from Steamed-Rice for High Performance Supercapacitors. *Materials Today Sustainability*, 2023, *24*. https://doi.org/10.1016/j.mtsust.2023.100480.

[28]   Chiu, Y. H.; Lin, L. Y. Effect of Activating Agents for Producing Activated Carbon Using a Facile One-Step Synthesis with Waste Coffee Grounds for Symmetric Supercapacitors. *J Taiwan Inst Chem Eng*, 2019, *101*, 177–185. https://doi.org/10.1016/j.jtice.2019.04.050.

[29]   Le Van, K.; Luong Thi Thu, T. Preparation of Pore-Size Controllable Activated Carbon from Rice Husk Using Dual Activating Agent and Its Application in Supercapacitor. *J Chem*, 2019, *2019*. https://doi.org/10.1155/2019/4329609.

[30]   Chen, W.; Gong, M.; Li, K.; Xia, M.; Chen, Z.; Xiao, H.; Fang, Y.; Chen, Y.; Yang, H.; Chen, H. Insight into KOH Activation Mechanism during Biomass Pyrolysis: Chemical Reactions between O-Containing Groups and KOH. *Appl Energy*, 2020, *278*, 115730. https://doi.org/https://doi.org/10.1016/j.apenergy.2020.115730.

[31]   Sumangala Devi, N.; Hariram, M.; Vivekanandhan, S. Modification Techniques to Improve the Capacitive Performance of Biocarbon Materials. *Journal of Energy Storage*. Elsevier Ltd January 1, 2021. https://doi.org/10.1016/j.est.2020.101870.

[32]   Navaneethan, D.; Krishna, S. K. Physicochemical Synthesis of Activated Carbon from Canna Indica (Biowaste) for High-Performance Supercapacitor Application. *Research on Chemical Intermediates*, 2023, *49* (4), 1387–1403. https://doi.org/10.1007/s11164-023-04955-2.

[33]   Gajalakshmi, T.; Kalaivani, T.; Thuy Lan Chi, N.; Brindhadevi, K. Investigation on Carbon Derived from Casuarina Bark Using Microwave Activation for High Performance Supercapacitors. *Fuel*, 2023, *337*. https://doi.org/10.1016/j.fuel.2022.127078.

[34]   Eleri, O. E.; Lou, F.; Yu, Z. Lithium-Ion Capacitors: A Review of Strategies toward Enhancing the Performance of the Activated Carbon Cathode. *Batteries*, 2023, *9* (11), 533. https://doi.org/10.3390/batteries9110533.

[35]   Xiaorui, L.; Haiping, Y. A State-of-the-Art Review of N Self-Doped Biochar Development in Supercapacitor Applications. *Frontiers in Energy Research*. Frontiers Media S.A. 2023. https://doi.org/10.3389/fenrg.2023.1135093.

[36]   Sangprasert, T.; Sattayarut, V.; Rajrujithong, C.; Khanchaitit, P.; Khemthong, P.; Chanthad, C.; Grisdanurak, N. Making Use of the Inherent Nitrogen Content of Spent Coffee Grounds to Create Nanostructured Activated Carbon for Supercapacitor and Lithium-Ion Battery Applications. *Diam Relat Mater*, 2022, *127*, 109164. https://doi.org/https://doi.org/10.1016/j.diamond.2022.109164.

[37]   Yao, S.; Zhang, Z.; Wang, Y.; Liu, Z.; Li, Z. Simple One-Pot Strategy for Converting Biowaste into Valuable Graphitized Hierarchically Porous Biochar for High-Efficiency Capacitive Storage. *J Energy Storage*, 2021, *44*, 103259. https://doi.org/https://doi.org/10.1016/j.est.2021.103259.

[38]   Wang, D.-W.; Li, F.; Yin, L.-C.; Lu, X.; Chen, Z.-G.; Gentle, I. R.; Lu, G. Q. (Max); Cheng, H.-M. Nitrogen-Doped Carbon Monolith for Alkaline Supercapacitors and Understanding Nitrogen-Induced Redox Transitions. *Chemistry – A European Journal*, 2012, *18* (17), 5345–5351. https://doi.org/https://doi.org/10.1002/chem.201102806.

[39]   Liang, X.; Liu, R.; Wu, X. Biomass Waste Derived Functionalized Hierarchical Porous Carbon with High Gravimetric and Volumetric Capacitances for Supercapacitors. *Microporous and Mesoporous Materials*, 2021, *310*, 110659. https://doi.org/https://doi.org/10.1016/j.micromeso.2020.110659.

[40]   Gao, Y.; Wang, J.; Huang, Y.; Zhang, S.; Zhang, S.; Zou, J. Rational Design of N-Doped Porous Biomass Carbon Nanofiber Electrodes for Flexible Asymmetric Supercapacitors with High-Performance. *Appl Surf Sci*, 2023, *638*. https://doi.org/10.1016/j.apsusc.2023.158137.

[41]   Yunita, A.; Farma, R.; Awitdrus, A.; Apriyani, I. The Effect of Various Electrolyte Solutions on the Electrochemical Properties of the Carbon Electrodes of Supercapacitor Cells Based on Biomass Waste. *Mater Today Proc*, 2023. https://doi.org/10.1016/j.matpr.2023.03.102.

[42]   Agrawal, A.; Gaur, A.; Kumar, A. Fabrication of Phyllanthus Emblica Leaves Derived High-Performance Activated Carbon-Based Symmetric Supercapacitor with Excellent Cyclic Stability. *J Energy Storage*, 2023, *66*. https://doi.org/10.1016/j.est.2023.107395.

[43]   Phukhrongthung, A.; Iamprasertkun, P.; Bunpheng, A.; Saisopa, T.; Umpuch, C.; Puchongkawarin, C.; Sawangphruk, M.; Luanwuthi, S. Oil Palm Leaf-Derived Hierarchical Porous Carbon for "Water-in-Salt" Based Supercapacitors: The Effect of Anions (Cl– and TFSI–) in Superconcentrated Conditions. *RSC Adv*, 2023, *13* (35), 24432–24444. https://doi.org/10.1039/d3ra03152g.

[44]   Mujtaba, Z.; Arshad, N. Moringa Oleifera and Its Spent Tea Waste Composites as Sustainable Anode Candidates for Energy Storage Applications: Morphological and Electrochemical Performance Studies. *Mater Chem Phys*, 2023, *301*. https://doi.org/10.1016/j.matchemphys.2023.127576.

[45]   Wang, B.; Li, Y.; Gu, Z.; Wang, H.; Liu, X.; Li, S.; Chen, X.; Liang, X.; Ogino, K.; Si, H. Biomass-Derived Hierarchically Porous Carbon-Supported MnOx Enable Fast Ion/Electron Transport at High Operating Voltages. *Fuel Processing Technology*, 2023, *249*. https://doi.org/10.1016/j.fuproc.2023.107857.

[46]   Inayat, A.; Albalawi, K.; Rehman, A. ur; Adnan; Saad, A. Y.; Saleh, E. A. M.; Alamri, M. A.; El-Zahhar, A. A.; Haider, A.; Abbas, S. M. Tunable Synthesis of Carbon Quantum Dots from the Biomass of Spent Tea Leaves as Supercapacitor Electrode. *Mater Today Commun*, 2023, *34*. https://doi.org/10.1016/j.mtcomm.2023.105479.

[47]   Kumari, R.; Kharangarh, P. R.; Singh, V.; Jha, R.; Ravi Kant, C. Sequential Processing of Nitrogen-Rich, Biowaste-Derived Carbon Quantum Dots Combined with Strontium Cobaltite for Enhanced Supercapacitive Performance. *J Alloys Compd*, 2023, *969*. https://doi.org/10.1016/j.jallcom.2023.172256.

[48]   Tyagi, A.; Mishra, K.; Shukla, V. K. Optimization of Sesamum Indicum Oil (Sesame Oil) Derived Activated Carbon Soot for Electric Double-Layer Capacitor (EDLC) Application. *Biomass Convers Biorefin*, 2023. https://doi.org/10.1007/s13399-023-04121-z.

[49]   Patil, A. V.; Sawant, S. A.; Sonkawade, R. G.; Vhatkar, R. S. Green Synthesized Carbon Aerogel for Electric Double Layer Capacitor. *J Energy Storage*, 2023, *72*. https://doi.org/10.1016/j.est.2023.108533.

[50]   Mohammed, A. A.; Panda, P. K.; Hota, A.; Tripathy, B. C.; Basu, S. Flexible Asymmetric Supercapacitor Based on Hyphaene Fruit Shell-Derived Multi-Heteroatom Doped Carbon and NiMoO4@NiCo2O4 Hybrid Structure Electrodes. *Biomass Bioenergy*, 2023, *179*. https://doi.org/10.1016/j.biombioe.2023.106981.

[51]   Said, B.; Bacha, O.; Rahmani, Y.; Harfouche, N.; Kheniche, H.; Zerrouki, D.; Belkhalfa, H.; Henni, A. Activated Carbon Prepared by Hydrothermal Pretreatment-Assisted Chemical Activation of Date Seeds for Supercapacitor Application. *Inorg Chem Commun*, 2023, *155*. https://doi.org/10.1016/j.inoche.2023.111012.

[52]   Yan, B.; Zhao, W.; Zhang, Q.; Kong, Q.; Chen, G.; Zhang, C.; Han, J.; Jiang, S.; He, S. One Stone for Four Birds: A "Chemical Blowing" Strategy to Synthesis Wood-Derived Carbon Monoliths for High-Mass Loading Capacitive Energy Storage in Low Temperature. *J Colloid Interface Sci*, 2024, *653*, 1526–1538. https://doi.org/10.1016/j.jcis.2023.09.179.

[53]   Yan, B.; Feng, L.; Zheng, J.; Zhang, Q.; Zhang, C.; Ding, Y.; Han, J.; Jiang, S.; He, S. In Situ Growth of N/O-Codoped Carbon Nanotubes in Wood-Derived Thick Carbon Scaffold to Boost the Capacitive Performance. *Colloids Surf A Physicochem Eng Asp*, 2023, *662*. https://doi.org/10.1016/j.colsurfa.2023.131018.

[54]   Pant, B.; Ojha, G. P.; Acharya, J.; Park, M. Preparation, Characterization, and Electrochemical Performances of Activated Carbon Derived from the Flower of Bauhinia Variegata L for Supercapacitor Applications. *Diam Relat Mater*, 2023, *136*. https://doi.org/10.1016/j.diamond.2023.110040.

[55]   Liu, C.; Yuan, R.; Yuan, Y.; Hou, R.; Liu, Y.; Ao, W.; Qu, J.; Yu, M.; Song, H.; Dai, J. An Environment-Friendly Strategy to Prepare Oxygen-Nitrogen-Sulfur Doped Mesopore-Dominant Porous Carbons for Symmetric Supercapacitors. *Fuel*, 2023, *344*. https://doi.org/10.1016/j.fuel.2023.128039.

[56]   Gnawali, C. L.; Shahi, S.; Manandhar, S.; Shrestha, G. K.; Adhikari, M. P.; Rajbhandari, R.; Pokharel, B. P. Porous Activated Carbon Materials from Triphala Seed Stones for High-Performance Supercapacitor Applications. *BIBECHANA*, 2023, *20* (1), 10–20. https://doi.org/10.3126/bibechana.v20i1.53432.

[57]   You, Z.; Zhao, L.; Zhao, K.; Liao, H.; Wen, S.; Xiao, Y.; Cheng, B.; Lei, S. Highly Tunable Three-Dimensional Porous Carbon Produced from Tea Seed Meal Crop by-Products for High Performance Supercapacitors. *Appl Surf Sci*, 2023, *607*. https://doi.org/10.1016/j.apsusc.2022.155080.

[58]   Zhang, Z.; Li, Y.; Yang, X.; Han, E.; Chen, G.; Yan, C.; Yang, X.; Zhou, D.; He, Y. In-Situ Confined Construction of N-Doped Compact Bamboo Charcoal Composites for Supercapacitors. *J Energy Storage*, 2023, *62*. https://doi.org/10.1016/j.est.2023.106954.

[59]   Zhang, Z.; Lu, S.; Li, Y.; Song, J.; Han, E.; Wang, H.; He, Y. Promoting Hierarchical Porous Carbon Derived from Bamboo via Copper Doping for High-Performance Supercapacitors. *Ind Crops Prod*, 2023, *203*. https://doi.org/10.1016/j.indcrop.2023.117155.

[60]   Qin, Z.; Ye, Y.; Zhang, D.; He, J.; Zhou, J.; Cai, J. One/Two-Step Contribution to Prepare Hierarchical Porous Carbon Derived from Rice Husk for Supercapacitor Electrode Materials. *ACS Omega*, 2023, *8* (5), 5088–5096. https://doi.org/10.1021/acsomega.2c07932.

[61]   Zhang, Z. W.; Lu, C. Y.; Liu, G. H.; Cao, Y. J.; Wang, Z.; Yang, T. ting; Kang, Y. H.; Wei, X. Y.; Bai, H. C. Self-Assembly of Caragana-Based Nanomaterials into Multiple Heteroatom-Doped 3D-Interconnected Porous Carbon for Advanced Supercapacitors. *Mater Today Adv*, 2023, *19*. https://doi.org/10.1016/j.mtadv.2023.100394.

[62]   Zhang, Z.; Qing, Y.; Wang, D.; Li, L.; Wu, Y. N-Doped Carbon Fibers Derived from Porous Wood Fibers Encapsulated in a Zeolitic Imidazolate Framework as an Electrode Material for Supercapacitors. *Molecules*, 2023, *28* (7). https://doi.org/10.3390/molecules28073081.

[63]   Zhao, Y.; Wang, Y.; Liu, Y.; Wang, H. Preparation of Hierarchical Porous Carbon through One-Step KOH Activation of Coconut Shell Biomass for High-Performance Supercapacitor. 2022. https://doi.org/10.21203/rs.3.rs-1895857/v1.

[64]   Li, Y.; Qi, B. Secondary Utilization of Jujube Shell Bio-Waste into Biomass Carbon for Supercapacitor Electrode Materials Study. *Electrochem commun*, 2023, *152*. https://doi.org/10.1016/j.elecom.2023.107512.

[65]   Bao, Y.; Xu, H.; Zhu, Y.; Chen, P.; Zhang, Y.; Chen, Y. 2,6-Diaminoanthraquinone Anchored on Functionalized Biomass Porous Carbon Boosts Electrochemical Stability for Metal-Free Redox Supercapacitor Electrode. *Electrochim Acta*, 2023, *437*. https://doi.org/10.1016/j.electacta.2022.141533.

[66]   Yao, Y.; Xie, T.; Li, P.; Du, W.; Jiang, J.; Ding, H.; Zhao, T.; Xu, G.; Zhang, L. Cheese-like Hierarchical Porous Carbon Material with Large Specific Surface Area Derived from Red Dates for High Performance Supercapacitors. *Diam Relat Mater*, 2023, *138*. https://doi.org/10.1016/j.diamond.2023.110169.

[67]   Hu, Y.; Ouyang, J.; Xiong, W.; Wang, R.; Lu, Y.; Yin, W.; Fan, Y.; Li, Z.; Du, K.; Li, X.; et al. A 3D Stacked Corrugated Pore Structure Composed of CoNiO2 and Polyaniline within the Tracheids of Wood-Derived Carbon for High-Performance Asymmetric Supercapacitors. *J Colloid Interface Sci*, 2023, *648*, 674–682. https://doi.org/10.1016/j.jcis.2023.05.191.

[68]   Geng, Y.; Wang, J.; Chen, X.; Wang, Q.; Zhang, S.; Tian, Y.; Liu, C.; Wang, L.; Wei, Z.; Cao, L.; et al. In Situ N, O-Dually Doped Nanoporous Biochar Derived from Waste Eutrophic Spirulina for High-Performance Supercapacitors. *Nanomaterials*, 2023, *13* (17). https://doi.org/10.3390/nano13172431.

[69]   Wang, S.; Zhou, J.; Zhang, Y.; He, S.; Esakkimuthu, S.; Zhu, K.; Kumar, S.; Lv, G.; Hu, X. Biochar Assisted Cultivation of Chlorella Protothecoides for Adsorption of Tetracycline and Electrochemical Study on Self-Cultured Chlorella Protothecoides. *Bioresour Technol*, 2023, *389*. https://doi.org/10.1016/j.biortech.2023.129810.

[70]   Çetin, R.; Arserim, M. A.; Akdemir, M. A Novel Study for Supercapacitor Applications via Corona Discharge Modified Activated Carbon Derived from Dunaliella Salina Microalgae. *J Energy Storage*, 2023, *72*. https://doi.org/10.1016/j.est.2023.108823.

[71]   Liu, Y.; Yang, Z.; He, J. Bacterial Assisted Sustainable Production of Three-Dimensional Defective Carbon from Rice Straw for Supercapacitor. *Mater Lett*, 2023, *352*. https://doi.org/10.1016/j.matlet.2023.135188.

[72]   Xia, Y.; Zuo, H.; Lv, J.; Wei, S.; Yao, Y.; Liu, Z.; Lin, Q.; Yu, Y.; Yu, W.; Huang, Y. Preparation of Multi-Layered Microcapsule-Shaped Activated Biomass Carbon with Ultrahigh Surface Area from Bamboo Parenchyma Cells for Energy Storage and Cationic Dyes Removal. *Journal of Cleaner Production*. Elsevier Ltd April 10, 2023. https://doi.org/10.1016/j.jclepro.2023.136517.

[73]   Nagaraju, Y. S.; Ganesh, H.; Veeresh, S.; Vijeth, H.; Devendrappa, H. A Strategy of Making Waste Profitable: Self-Templated Synthesis of Helical Rod Structured Porous Carbon

Derived from Ganoderma Lucidem for Advanced Supercapacitor Electrode. *Diam Relat Mater*, 2023, *131*. https://doi.org/10.1016/j.diamond.2022.109607.

[74]   Zou, X.; Dong, C.; Jin, Y.; Wang, D.; Li, L.; Wu, S.; Xu, Z.; Chen, Y.; Li, Z.; Yang, H. Engineering of N, P Co-Doped Hierarchical Porous Carbon from Sugarcane Bagasse for High-Performance Supercapacitors and Sodium Ion Batteries. *Colloids Surf A Physicochem Eng Asp*, 2023, *672*. https://doi.org/10.1016/j.colsurfa.2023.131715.

[75]   He, X.; Li, W.; Xia, Y.; Huang, H.; Xia, X.; Gan, Y.; Zhang, J.; Zhang, W. Pivotal Factors of Wood-Derived Electrode for Supercapacitor: Component Striping, Specific Surface Area and Functional Group at Surface. *Carbon N Y*, 2023, *210*. https://doi.org/10.1016/j.carbon.2023.118090.

[76]   Du, X.; Lin, Z.; Zhang, Y.; Li, P. Microstructural Tailoring of Porous Few-Layer Graphene-like Biochar from Kitchen Waste Hydrolysis Residue in Molten Carbonate Medium: Structural Evolution and Conductive Additive-Free Supercapacitor Application. *Science of the Total Environment*, 2023, *871*. https://doi.org/10.1016/j.scitotenv.2023.162045.

[77]   Xuan, X.; Wang, M.; You, W.; Manickam, S.; Tao, Y.; Yoon, J. Y.; Sun, X. Hydrodynamic Cavitation-Assisted Preparation of Porous Carbon from Garlic Peels for Supercapacitors. *Ultrason Sonochem*, 2023, *94*. https://doi.org/10.1016/j.ultsonch.2023.106333.

[78]   Tian, X. L.; Yu, J. H.; Qiu, L.; Zhu, Y. H.; Zhu, M. Q. Structural Changes and Electrochemical Properties of Mesoporous Activated Carbon Derived from Eucommia Ulmoides Wood Tar by KOH Activation for Supercapacitor Applications. *Ind Crops Prod*, 2023, *197*. https://doi.org/10.1016/j.indcrop.2023.116628.

[79]   Meng, X.; Zhang, D.; Wang, B.; He, Y.; Xia, X.; Yang, B.; Han, Z. Biomass-Derived Phosphorus-Doped Hierarchical Porous Carbon Fabricated by Microwave Irritation under Ambient Atmosphere with High Supercapacitance Performance in Trifluoroacetic Acid Electrolyte. *J Energy Storage*, 2023, *57*. https://doi.org/10.1016/j.est.2022.106345.

[80]   Wang, X.; Wang, Y.; Yan, L.; Wang, Q.; Li, J.; Zhong, X.; Liu, Q.; Li, Q.; Cui, S.; Xie, G. From Pollutant to High-Performance Supercapacitor: Semi-Coking Wastewater Derived N–O–S Self-Doped Porous Carbon. *Colloids Surf A Physicochem Eng Asp*, 2023, *657*. https://doi.org/10.1016/j.colsurfa.2022.130596.

[81]   Chen, X.; Zhao, C.; Ma, J.; Sun, X. Multifunctional Porous Carbon Materials with Excellent Supercapacitor Performance and Oxygen Reduction Properties Derived from Pomelo Peel.

[82]   Ma, C.; Li, W.; Cheng, H.; Li, Z.; He, G. Biochar for Supercapacitor Electrodes: Mechanisms in Aqueous Electrolytes.

[83]   Yao, W.; Zheng, D.; Li, Z.; Wang, Y.; Tan, H.; Zhang, Y. MXene@ Carbonized Wood Monolithic Electrode with Hierarchical Porous Framework for High-Performance Supercapacitors. *Appl Surf Sci*, 2023, *638*. https://doi.org/10.1016/j.apsusc.2023.158130.

[84]   Wu, W.; Zheng, H.; Zhang, Y.; Wang, Q.; Huang, W.; Xiang, J.; Yang, X.; Lu, W.; Zhang, Z.; Wang, S. Preparation of High Performance Supercapacitors with Nitrogen and Oxygen Self-Doped Porous Carbon Derived from Wintersweet-Fruit-Shell. *Journal of Physics and Chemistry of Solids*, 2023, *177*. https://doi.org/10.1016/j.jpcs.2023.111274.

[85]   Tian, W.; Ren, P.; Wang, J.; Hou, X.; Sun, A.; Jin, Y.; Chen, Z. Facile Synthesis of Dense Porous Carbon Derived from Linum Usitatissimum L. Root for High Mass Loading Supercapacitors. *J Energy Storage*, 2023, *63*. https://doi.org/10.1016/j.est.2023.107039.

[86]   Tian, W.; Ren, P.; Hou, X.; Guo, Z.; Xue, R.; Chen, Z.; Jin, Y. Biomass Derived N/O Self-Doped Porous Carbon for Advanced Supercapacitor Electrodes. *Ind Crops Prod*, 2023, *202*. https://doi.org/10.1016/j.indcrop.2023.117032.

[87]   Zhu, W.; Shen, D.; Xie, H. Combination of Chemical Activation and Nitrogen Doping toward Hierarchical Porous Carbon from Houttuynia Cordata for Supercapacitors. *J Energy Storage*, 2023, *60*. https://doi.org/10.1016/j.est.2022.106595.

[88]   Goodwin, V.; Jitreewas, P.; Sesuk, T.; Limthongkul, P.; Charojrochkul, S. Development of Modified Mesoporous Carbon from Palm Oil Biomass for Energy Storage Supercapacitor Application. In *IOP Conference Series: Earth and Environmental Science*; Institute of Physics, 2023; Vol. 1199. https://doi.org/10.1088/1755-1315/1199/1/012003.

[89]   Wang, T.; Peng, L.; Wu, D.; Chen, B.; Jia, B. Crude Fiber and Protein Rich Cottonseed Meal Derived Carbon Quantum Dots Composite Porous Carbon for Supercapacitor. *J Alloys Compd*, 2023, *947*. https://doi.org/10.1016/j.jallcom.2023.169499.

[90]   Meekati, T.; Pakawatpanurut, P.; Amornsakchai, T.; Meethong, N.; Sodtipinta, J. High-Performance Activated Carbon Fiber Derived from Pineapple Leaf Fiber via Co(II)-Assisted Preparation Revealed Surprising Capacitive-Pseudocapacitive Synergistic Charge Storage. *J Energy Storage*, 2023, *71*. https://doi.org/10.1016/j.est.2023.108135.

[91]   Nguyen, T. B.; Yoon, B.; Nguyen, T. D.; Oh, E.; Ma, Y.; Wang, M.; Suhr, J. A Facile Salt-Templating Synthesis Route of Bamboo-Derived Hierarchical Porous Carbon for Supercapacitor Applications. *Carbon N Y*, 2023, *206*, 383–391. https://doi.org/10.1016/j.carbon.2023.02.060.

[92]   Salheen, S. A.; Nassar, H. F.; Dsoke, S.; El-Deen, A. G. Constructing Ultraporous Activated Hollow Carbon Nanospheres Derived from Rotten Grapes for Boosting Energy Density and Lifespan Supercapacitors. *Colloids Surf A Physicochem Eng Asp*, 2023, *660*. https://doi.org/10.1016/j.colsurfa.2022.130821.

[93]   Rajasekaran, S. J.; Grace, A. N.; Jacob, G.; Alodhayb, A.; Pandiaraj, S.; Raghavan, V. Investigation of Different Aqueous Electrolytes for Biomass-Derived Activated Carbon-Based Supercapacitors. *Catalysts*, 2023, *13* (2). https://doi.org/10.3390/catal13020286.

[94]   Bhosale, S. R.; Bhosale, R. R.; Shinde, S. B.; Moyo, A. A.; Dhavale, R. P.; Kolekar, S. S.; Anbhule, P. V. Biomass Nanoarchitectonics with Annona Reticulate Flowers for Mesoporous Carbon Impregnated on CeO2 Nanogranular Electrode for Energy Storage Application. *Inorg Chem Commun*, 2023, *154*. https://doi.org/10.1016/j.inoche.2023.110949.

[95]   Siva Priya, D.; Kennedy, L. J.; Anand, G. T. Engineering the Pongamia Pinnata Oil Pressed Residue Derived Porous Carbon for Symmetric Supercapacitor Application.

[96]   Qin, S.; Liu, P.; Wang, J.; Liu, C.; Zhang, S.; Tian, Y.; Zhang, F.; Wang, L.; Cao, L.; Zhang, J.; et al. In Situ N, O Co-Doped Nanoporous Carbon Derived from Mixed Egg and Rice Waste as Green Supercapacitor. *Molecules*, 2023, *28* (18). https://doi.org/10.3390/molecules28186543.

[97]   Bhat, S.; Uthappa, U. T.; Sadhasivam, T.; Altalhi, T.; Soo Han, S.; Kurkuri, M. D. Abundant Cilantro Derived High Surface Area Activated Carbon (AC) for Superior Adsorption Performances of Cationic/Anionic Dyes and Supercapacitor Application. *Chemical Engineering Journal*, 2023, *459*. https://doi.org/10.1016/j.cej.2023.141577.

[98]   Rawat, S.; Boobalan, T.; Sathish, M.; Hotha, S.; Thallada, B. Utilization of CO2 Activated Litchi Seed Biochar for the Fabrication of Supercapacitor Electrodes. *Biomass Bioenergy*, 2023, *171*. https://doi.org/10.1016/j.biombioe.2023.106747.

[99] Jiao, S.; Yao, Y.; Zhang, J.; Zhang, L.; Li, C.; Zhang, H.; Zhao, X.; Chen, H.; Jiang, J. Nano-Flower-like Porous Carbon Derived from Soybean Straw for Efficient N-S Co-Doped Supercapacitors by Coupling in-Situ Heteroatom Doping with Green Activation Method. *Appl Surf Sci*, 2023, *615*. https://doi.org/10.1016/j.apsusc.2023.156365.

[100] Ma, R.; Luo, W.; Yan, L.; Guo, C.; Ding, X.; Gong, X.; Jia, D.; Xu, M.; Ai, L.; Guo, N.; et al. Constructing the Quinonyl Groups and Structural Defects in Carbon for Supercapacitor and Capacitive Deionization Applications. *J Colloid Interface Sci*, 2023, *645*, 685–693. https://doi.org/10.1016/j.jcis.2023.04.119.

[101] Jiang, R.; Zhou, C.; Yang, Y.; Zhu, S.; Li, S.; Zhou, J.; Li, W.; Ding, L. Rice Straw-Derived Activated Carbon/Nickel Cobalt Sulfide Composite for High Performance Asymmetric Supercapacitor. *Diam Relat Mater*, 2023, *139*. https://doi.org/10.1016/j.diamond.2023.110322.

[102] Mehdi, R.; Naqvi, S. R.; Khoja, A. H.; Hussain, R. Biomass Derived Activated Carbon by Chemical Surface Modification as a Source of Clean Energy for Supercapacitor Application. *Fuel*, 2023, *348*. https://doi.org/10.1016/j.fuel.2023.128529.

[103] Dubey, P.; Maheshwari, P. H.; Mansi; Shrivastav, V.; Sundriyal, S. Effect of Nitrogen and Sulphur Co-Doping on the Surface and Diffusion Characteristics of Date Seed-Derived Porous Carbon for Asymmetric Supercapacitors. *J Energy Storage*, 2023, *58*. https://doi.org/10.1016/j.est.2022.106441.

[104] Atchudan, R.; Perumal, S.; Sundramoorthy, A. K.; Manoj, D.; Kumar, R. S.; Almansour, A. I.; Lee, Y. R. Facile Synthesis of Functionalized Porous Carbon by Direct Pyrolysis of Anacardium Occidentale Nut-Skin Waste and Its Utilization towards Supercapacitors. *Nanomaterials*, 2023, *13* (10). https://doi.org/10.3390/nano13101654.

[105] Qin, Q.; Zhong, F.; Song, T.; Yang, Z.; Zhang, P.; Cao, H.; Niu, W.; Yao, Z. Optimization of Multiscale Structure and Electrochemical Properties of Bamboo-Based Porous Activated Biochar by Coordinated Regulation of Activation and Air Oxidation. *Chemical Engineering Journal*, 2023, *477*. https://doi.org/10.1016/j.cej.2023.146763.

[106] Numee, P.; Sangtawesin, T.; Yilmaz, M.; Kanjana, K. Activated Carbon Derived from Radiation-Processed Durian Shell for Energy Storage Application. *Carbon Resources Conversion*, 2023. https://doi.org/10.1016/j.crcon.2023.07.001.

[107] Qi, P.; Su, Y.; Yang, L.; Wang, J.; Jiang, M.; Sun, X.; Zhang, P.; Xiong, Y. Nanoarchitectonics of Hierarchical Porous Carbon Based on Carbonization of Heavy Fraction of Bio-Oil and Its Supercapacitor Performance. *J Energy Storage*, 2023, *74*. https://doi.org/10.1016/j.est.2023.109398.

[108] Ozpinar, P.; Dogan, C.; Demiral, H.; Morali, U.; Erol, S.; Yildiz, D.; Samdan, C.; Demiral, I. Effect of Binder on the Electrochemical Performance of Activated Carbon Electrodes Obtained from Waste Hazelnut Shells: Comparison of PTFE and PVDF. *Diam Relat Mater*, 2023, *137*. https://doi.org/10.1016/j.diamond.2023.110092.

[109] Benwannamas, N.; Sangtawesin, T.; Yilmaz, M.; Kanjana, K. Gamma-Induced Interconnected Networks in Microporous Activated Carbons from Palm Petiole under NaNO3 Oxidizing Environment towards High-Performance Electric Double Layer Capacitors (EDLCs). *Sci Rep*, 2023, *13* (1). https://doi.org/10.1038/s41598-023-40176-8.

[110] Kumaresan, N.; Alsalhi, M. S.; Karuppasamy, P.; Praveen Kumar, M.; Pandian, M. S.; Arulraj, A.; Peera, S. G.; Mangalaraja, R. V.; Devanesan, S.; Ramasamy, P.; et al. Nitrogen Implanted Carbon Nanosheets Derived from Acorus Calamus as an Efficient Electrode for the

Supercapacitor Application. *Molecular Catalysis*, 2023, *538*. https://doi.org/10.1016/j.mcat.2023.112978.

[111] Karakehya, N. Effects of One-Step and Two-Step KOH Activation Method on the Properties and Supercapacitor Performance of Highly Porous Activated Carbons Prepared from Lycopodium Clavatum Spores. *Diam Relat Mater*, 2023, *135*. https://doi.org/10.1016/j.diamond.2023.109873.

[112] Gouda, M. S.; Shehab, M.; Helmy, S.; Soliman, M.; Salama, R. S. Nickel and Cobalt Oxides Supported on Activated Carbon Derived from Willow Catkin for Efficient Supercapacitor Electrode. *J Energy Storage*, 2023, *61*. https://doi.org/10.1016/j.est.2023.106806.

[113] Gouda, M. S.; Shehab, M.; Soliman, M. M.; Helmy, S.; Salama, R. S. Preparation and Characterization of Supercapacitor Electrodes Utilizing Catkin Plant as an Activated Carbon Source; 2023.

[114] Pallavolu, M. R.; Prabhu, S.; Nallapureddy, R. R.; Kumar, A. S.; Banerjee, A. N.; Joo, S. W. Bio-Derived Graphitic Carbon Quantum Dot Encapsulated S- and N-Doped Graphene Sheets with Unusual Battery-Type Behavior for High-Performance Supercapacitor. *Carbon N Y*, 2023, *202*, 93–102. https://doi.org/10.1016/j.carbon.2022.10.077.

[115] Jalalah, M.; Han, H. S.; Mahadani, M.; Nayak, A. K.; Harraz, F. A. Novel Interconnected Hierarchical Porous Carbon Derived from Biomass for Enhanced Supercapacitor Application. *Journal of Electroanalytical Chemistry*, 2023, *935*. https://doi.org/10.1016/j.jelechem.2023.117355.

[116] Wang, M. xi; He, D.; Zhu, M.; Wu, L.; Wang, Z.; Huang, Z. H.; Yang, H. Green Fabrication of Hierarchically Porous Carbon Microtubes from Biomass Waste via Self-Activation for High-Energy-Density Supercapacitor. *J Power Sources*, 2023, *560*. https://doi.org/10.1016/j.jpowsour.2023.232703.

[117] Merin, P.; Jimmy Joy, P.; Muralidharan, M. N.; Veena Gopalan, E.; Seema, A. Biomass-Derived Activated Carbon for High-Performance Supercapacitor Electrode Applications. *Chem Eng Technol*, 2021, *44* (5), 844–851. https://doi.org/10.1002/ceat.202000450.

[118] Ma, M.; Zhang, J.; Huan, Y.; Ren, M.; Wei, T.; Yan, S. 3D Stack Tubular Mesoporous Carbon Derived from Discarded Sesame Capsule Shells for High-Performance Supercapacitors. *Diam Relat Mater*, 2023, *131*. https://doi.org/10.1016/j.diamond.2022.109562.

[119] J.E., M.; Chandewar, P. R.; Shee, D.; Sankar Mal, S. Phosphomolybdic Acid Embedded into Biomass-Derived Biochar Carbon Electrode for Supercapacitor Applications. *Journal of Electroanalytical Chemistry*, 2023, *936*. https://doi.org/10.1016/j.jelechem.2023.117354.

[120] Liu, D.; Xu, G.; Yuan, X.; Ding, Y.; Fan, B. Pore Size Distribution Modulation of Waste Cotton-Derived Carbon Materials via Citrate Activator to Boost Supercapacitive Performance. *Fuel*, 2023, *332*. https://doi.org/10.1016/j.fuel.2022.126044.

[121] Lei, D.; Zeng, Y.; Zhong, J.; Chen, J.; Ye, Y.; Wang, W. Ultra-High Specific Surface Area Porous Carbons Derived from Chinese Medicinal Herbal Residues with Potential Applications in Supercapacitors and CO_2 Capture. *Colloids Surf A Physicochem Eng Asp*, 2023, *666*. https://doi.org/10.1016/j.colsurfa.2023.131327.

[122] Bejjanki, D.; Banothu, P.; Kumar, V. B.; Kumar, P. S. Biomass-Derived N-Doped Activated Carbon from Eucalyptus Leaves as an Efficient Supercapacitor Electrode Material. *C-Journal of Carbon Research*, 2023, *9* (1). https://doi.org/10.3390/c9010024.

[123] Zeng, F.; Meng, Z.; Xu, Z.; Xu, J.; Shi, W.; Wang, H.; Hu, X.; Tian, H. Biomass-Derived Porous Activated Carbon for Ultra-High Performance Supercapacitor Applications and High Flux Removal of Pollutants from Water. *Ceram Int*, 2023, *49* (10), 15377–15386. https://doi.org/10.1016/j.ceramint.2023.01.122.

[124] Abdallah, F.; Arthur, E. K.; Mensah-Darkwa, K.; Gikunoo, E.; Baffour, S. A.; Agamah, B. A.; Nartey, M. A.; Agyemang, F. O. Electrochemical Performance of Corncob-Derived Activated Carbon-Graphene Oxide and TiO2 Ternary Composite Electrode for Supercapacitor Applications. *J Energy Storage*, 2023, *68*. https://doi.org/10.1016/j.est.2023.107776.

[125] Deng, F.; Li, Y.; Zhang, Y.; Zhang, Q.; Li, Y.; Shang, J.; Wang, J.; Gao, R.; Li, R. A Popcorn-Derived Porous Carbon Optimized by Thermal Treatment and Its Outstanding Electrochemical Performance. *J Energy Storage*, 2023, *60*. https://doi.org/10.1016/j.est.2023.106668.

[126] Darjazi, H.; Bottoni, L.; Moazami, H. R.; Rezvani, S. J.; Balducci, L.; Sbrascini, L.; Staffolani, A.; Tombesi, A.; Nobili, F. From Waste to Resources: Transforming Olive Leaves to Hard Carbon as Sustainable and Versatile Electrode Material for Li/Na-Ion Batteries and Supercapacitors. *Materials Today Sustainability*, 2023, *21*. https://doi.org/10.1016/j.mtsust.2022.100313.

[127] Zhang, P.; Li, Y.; Wang, M.; Zhang, D.; Ouyang, W.; Liu, L.; Wang, M.; Zhang, K.; Wang, H.; Chen, C. Self-Doped (N/O/S) Nanoarchitectonics of Hierarchically Porous Carbon from Palm Flower for High-Performance Supercapacitors. *Diam Relat Mater*, 2023, *136*. https://doi.org/10.1016/j.diamond.2023.109976.

[128] Selvaraj, M.; Balamoorthy, E.; Manivasagam, T. G. Biomass Derived Nitrogen-Doped Activated Carbon and Novel Biocompatible Gel Electrolytes for Solid-State Supercapacitor Applications. *J Energy Storage*, 2023, *72*. https://doi.org/10.1016/j.est.2023.108543.

[129] Tamasauskaite-Tamasiunaite, L.; Jablonskienė, J.; Šimkūnaitė, D.; Volperts, A.; Plavniece, A.; Dobele, G.; Zhurinsh, A.; Jasulaitiene, V.; Niaura, G.; Drabavicius, A.; et al. Black Liquor and Wood Char-Derived Nitrogen-Doped Carbon Materials for Supercapacitors. *Materials*, 2023, *16* (7). https://doi.org/10.3390/ma16072551.

[130] Qiu, L.; Guo, R.; Ma, X.; Li, J. Fabrication of Hierarchical Porous Biomass-Based Carbon Aerogels from Liqueed Wood for Supercapacitor Applications. 2023. https://doi.org/10.21203/rs.3.rs-2596680/v1.

[131] Feng, L.; Yan, B.; Zheng, J.; Zhang, Q.; Wei, R.; Zhang, C.; Han, J.; Jiang, S.; He, S. Chemical Foaming-Assisted Synthesis of N, O Co-Doped Hierarchical Porous Carbon from Soybean Protein for High Rate Performance Supercapacitors. *Diam Relat Mater*, 2023, *133*. https://doi.org/10.1016/j.diamond.2023.109767.

[132] Li, K.; Liu, Z.; Ma, X.; Feng, Q.; Wang, D.; Ma, D. A Combination of Heteroatom Doping Engineering Assisted by Molten Salt and KOH Activation to Obtain N and O Co-Doped Biomass Porous Carbon for High Performance Supercapacitors. *J Alloys Compd*, 2023, *960*. https://doi.org/10.1016/j.jallcom.2023.170785.

[133] Liang, K.; Chen, Y.; Wang, S.; Wang, D.; Wang, W.; Jia, S.; Mitsuzakic, N.; Chen, Z. Peanut Shell Waste Derived Porous Carbon for High-Performance Supercapacitors. *J Energy Storage*, 2023, *70*. https://doi.org/10.1016/j.est.2023.107947.

[134] Januszewicz, K.; Kazimierski, P.; Cymann-Sachajdak, A.; Hercel, P.; Barczak, B.; Wilamowska-Zawłocka, M.; Kardaś, D.; Łuczak, J. Conversion of Waste Biomass to Designed and Tailored Activated Chars with Valuable Properties for Adsorption and

Electrochemical Applications. *Environmental Science and Pollution Research*, 2023, *30* (43), 96977–96992. https://doi.org/10.1007/s11356-023-28824-y.

[135] Manoharan, K.; Sundaram, R.; Raman, K. Preparation and Characterization of Hydrogen Storage Medium (IMO/TPAC) and Asymmetric Supercapacitor (IMO/TPAC‖TPAC) Using Imogolite (IMO) and Biomass Derived Activated Carbon from Tangerine Peel (TPAC) for Renewable Energy Storage Applications. *Int J Hydrogen Energy*, 2023, *48* (74), 28694–28711. https://doi.org/10.1016/j.ijhydene.2023.04.076.

[136] Vignesh, K.; Ganeshbabu, M.; Puneeth, N. P. N.; Mathivanan, T.; Ramkumar, B.; Lee, Y. S.; Selvan, R. K. Oxygen-Rich Functionalized Porous Carbon by KMnO4 Activation on Pods of Prosopis Juliflora for Symmetric Supercapacitors. *J Energy Storage*, 2023, *72*. https://doi.org/10.1016/j.est.2023.108216.

[137] Vengadesan, K.; Madaswamy, S. L.; Natarajan, V. K.; Dhanusuraman, R. Biocarbon Derived from Seeds of Palmyra Palm Tree for a Supercapacitor Application. *Advanced Nano Research*, 2023, *6* (1), 1–10. https://doi.org/10.21467/anr.6.1.1-10.

[138] Zhang, J.; Jiao, S.; Luo, H.; Gao, M.; Li, C.; Zhang, H.; Kong, F.; Zhao, X.; Chen, H.; Jiang, J. Fabrication of 3D Self-Supporting Hierarchical Poplar-Based Thick Carbon Electrode via Coupling Delignification and Gradient Carbonization for Electrochemical Performance. *J Power Sources*, 2023, *562*. https://doi.org/10.1016/j.jpowsour.2023.232787.

[139] Jia, J.; Yao, Z.; Zhao, L.; Xie, T.; Sun, Y.; Tian, L.; Huo, L.; Liu, Z. Functionalization of Supercapacitors Electrodes Oriented Hydrochar from Cornstalk: A New Vision via Biomass Fraction. *Biomass Bioenergy*, 2023, *175*. https://doi.org/10.1016/j.biombioe.2023.106858.

[140] Tu, J.; Qiao, Z.; Wang, Y.; Li, G.; Zhang, X.; Li, G. American Ginseng Biowaste-Derived Activated Carbon for High-Performance Supercapacitors. *Int J Electrochem Sci*, 2023, *18* (2), 16–24. https://doi.org/10.1016/j.ijoes.2023.01.011.

[141] Liao, H.; Zhong, L.; Deng, Y.; Chen, H.; Liao, G.; Xiao, Y.; Cheng, B.; Lei, S. A Systematic Study on Equisetum Ramosissimum Desf. Derived Honeycomb Porous Carbon for Supercapacitors: Insight into the Preparation-Structure-Performance Relationship. *Appl Surf Sci*, 2023, *623*. https://doi.org/10.1016/j.apsusc.2023.157010.

[142] Liu, H.; Huang, X.; Zhou, M.; Gu, J.; Xu, M.; Jiang, L.; Zheng, M.; Li, S.; Miao, Z. Efficient Conversion of Biomass Waste to N/O Co-Doped Hierarchical Porous Carbon for High Performance Supercapacitors. *J Anal Appl Pyrolysis*, 2023, *169*. https://doi.org/10.1016/j.jaap.2022.105844.

[143] Su, H.; Lan, C.; Wang, Z.; Zhu, L.; Zhu, M. Controllable Preparation of Eucommia Wood-Derived Mesoporous Activated Carbon as Electrode Materials for Supercapacitors. *Polymers (Basel)*, 2023, *15* (3). https://doi.org/10.3390/polym15030663.

[144] Liu, H.; Wang, Y.; Lv, L.; Liu, X.; Wang, Z.; Liu, J. Oxygen-Enriched Hierarchical Porous Carbons Derived from Lignite for High-Performance Supercapacitors. *Energy*, 2023, *269*. https://doi.org/10.1016/j.energy.2023.126707.

[145] Inal, I. I. G.; Koyuncu, F.; Güzel, F. Investigating the Surface Properties of Red Pepper Industrial Waste-Based Activated Carbons for Use as Reversible Supercapacitor Electrodes. *Diam Relat Mater*, 2023, *138*. https://doi.org/10.1016/j.diamond.2023.110212.

[146] Li, K.; Nan, D. hong; Li, Z. yu; Xie, J. heng; Ma, S. wei; Huang, Y. qin; Lu, Q. Preparation and Optimization of Nitrogen-Doped Porous Carbon Derived from Bio-Oil Distillation Residue for High-Performance Supercapacitors. *J Energy Storage*, 2023, *57*. https://doi.org/10.1016/j.est.2022.106219.

[147] Li, L.; Wei, X. Y.; Shao, C. W.; Yin, F.; Sun, B. K.; Liu, F. J.; Li, J. H.; Liu, Z. Q.; Zong, Z. M. Honeycomb-like N/O Self-Doped Hierarchical Porous Carbons Derived from Low-Rank Coal and Its Derivatives for High-Performance Supercapacitor. *Fuel*, 2022, *331*. https://doi.org/10.1016/j.fuel.2022.125658.

[148] Lin, L.; Zheng, Z.; Li, X.; Park, S.; Zhang, W.; Diao, G.; Piao, Y. Design Strategy for Porous Carbon Nanomaterials from Rational Utilization of Natural Rubber Latex Foam Scraps. *Ind Crops Prod*, 2023, *192*. https://doi.org/10.1016/j.indcrop.2022.116036.

[149] Zhou, Q.; Li, H.; Jia, B.; Dang, Y.; Zhang, G. One-Pot Synthesis of Porous Carbon from Chinese Medicine Residues Driven by Potassium Citrate and Application in Supercapacitors. *J Anal Appl Pyrolysis*, 2023, *170*. https://doi.org/10.1016/j.jaap.2023.105894.

[150] Xu, Q.; Ni, X.; Chen, S.; Ye, J.; Yang, J.; Wang, H.; Li, D.; Yuan, H. Hierarchically Porous Carbon from Biomass Tar as Sustainable Electrode Material for High-Performance Supercapacitors. *Int J Hydrogen Energy*, 2023, *48* (66), 25635–25644. https://doi.org/10.1016/j.ijhydene.2023.03.171.

[151] Qin, S.; Liu, P.; Wang, J.; Liu, C.; Wang, Q.; Chen, X.; Zhang, S.; Tian, Y.; Zhang, F.; Wang, L.; et al. In Situ N, O Co-Doped Porous Carbon Derived from Antibiotic Fermentation Residues as Electrode Material for High-Performance Supercapacitors. *RSC Adv*, 2023, *13* (34), 24140–24149. https://doi.org/10.1039/d3ra04164f.

[152] Meng, X.; Wang, X.; Li, W.; Kong, F.; Zhang, F. Fabrication of N-Doped Porous Carbon with Micro/Mesoporous Structure from Furfural Residue for Supercapacitors. *Polymers (Basel)*, 2023, *15* (19). https://doi.org/10.3390/polym15193976.

[153] Li, Y.; Jia, X.; Li, X.; Liu, P.; Zhang, X.; Guo, M. Study on the Potential of Sludge-Derived Humic Acid as Energy Storage Material. *Waste Management*, 2023, *162*, 55–62. https://doi.org/10.1016/j.wasman.2023.03.015.

[154] Bian, Z.; Wang, H.; Zhao, X.; Ni, Z.; Zhao, G.; Chen, C.; Hu, G.; Komarneni, S. Optimized Mesopores Enable Enhanced Capacitance of Electrochemical Capacitors Using Ultrahigh Surface Area Carbon Derived from Waste Feathers. *J Colloid Interface Sci*, 2023, *630*, 115–126. https://doi.org/10.1016/j.jcis.2022.09.123.

[155] Xue, B.; Xu, J.; Chen, Z.; Xiao, R. Valorizing High-Fraction Bio-Oil to Prepare 3D Interconnected Porous Carbon with Efficient Pore Utilization for Supercapacitor Applications. *Fuel Processing Technology*, 2023, *239*. https://doi.org/10.1016/j.fuproc.2022.107538.

[156] Xue, B.; Xu, J.; Xiao, R. Ice Template-Assisting Activation Strategy to Prepare Biomass-Derived Porous Carbon Cages for High-Performance Zn-Ion Hybrid Supercapacitors. *Chemical Engineering Journal*, 2023, *454*. https://doi.org/10.1016/j.cej.2022.140192.

[157] Liu, B.; Wang, X.; Chen, Y.; Xie, H.; Zhao, X.; Nassr, A. B.; Li, Y. Honeycomb Carbon Obtained from Coal Liquefaction Residual Asphaltene for High-Performance Supercapacitors in Ionic and Organic Liquid-Based Electrolytes. *J Energy Storage*, 2023, *68*. https://doi.org/10.1016/j.est.2023.107826.

[158] Hou, Y.; Sun, Q.; Du, H. Fabrication of Nickel Sulfide/Hierarchical Porous Carbon from Refining Waste for High-Performance Supercapacitor. *J Alloys Compd*, 2023, *937*. https://doi.org/10.1016/j.jallcom.2022.168399.

[159] Gokce, Y. Waste Jean Derived Self N-Containing Activated Carbon as a Potential Electrode Material for Supercapacitors. *Turk J Chem*, 2023, *47* (4), 789–800. https://doi.org/10.55730/1300-0527.3579.

[160] Chen, Y.; Qin, F.; Wang, Z.; Chen, S.; Cao, Y.; Zhang, S.; Huang, X.; Li, Y. Dense Porous Carbon from Chemical Welding the Oxidized Coal Liquefaction Residue for Enhanced Volumetric Performance Supercapacitors. *J Energy Storage*, 2023, *72*. https://doi.org/10.1016/j.est.2023.108542.

[161] Wang, X.; Liu, B.; Wang, S.; Xie, H.; Zha, Y.; Huang, X.; Santos, D. M. F.; Li, Y. Oxygen Self-Doped Hierarchical Porous Carbons Derived from Coal Liquefaction Residue for High-Performance Supercapacitors in Organic and Ionic Liquid-Based Electrolytes. *Colloids Surf A Physicochem Eng Asp*, 2023, *669*. https://doi.org/10.1016/j.colsurfa.2023.131552.

[162] Brandão, A. T. S. C.; State, S.; Costa, R.; Potorac, P.; Vázquez, J. A.; Valcarcel, J.; Silva, A. F.; Anicai, L.; Enachescu, M.; Pereira, C. M. Renewable Carbon Materials as Electrodes for High-Performance Supercapacitors: From Marine Biowaste to High Specific Surface Area Porous Biocarbons. *ACS Omega*, 2023, *8* (21), 18782–18798. https://doi.org/10.1021/acsomega.3c00816.

[163] Brandão, A. T. S. C.; Costa, R.; State, S.; Potorac, P.; Dias, C.; Vázquez, J. A.; Valcarcel, J.; Silva, A. F.; Enachescu, M.; Pereira, C. M. Chitins from Seafood Waste as Sustainable Porous Carbon Precursors for the Development of Eco-Friendly Supercapacitors. *Materials*, 2023, *16* (6). https://doi.org/10.3390/ma16062332.

Emerging Materials for Next Frontier Energy and Environment Applications   Materials Research Forum LLC
Materials Research Foundations 170 (2024) 94-116   https://doi.org/10.21741/9781644903292-5

Chapter 5

Prospects of metal organic framework (MOF) as excellent electrode material for supercapacitors

Ariharaputhiran Anitha[1,a], Amalraj John[2,b]

[1]Assistant Professor of Chemistry, V.V.Vanniaperumal College for Women, Virudhunagar, India 626001

[2]Assistant Professor, Institute of Natural Resources Chemistry, University of Talca, Chile.

[a]anithaa@vvvcollege.org, [b]jamalraj@utalca.cl

Abstract

Supercapacitors(SCs) are the need of the hour as it has salient features like fast charging/discharging and long cycle life. As the electrode plays a crucial role of its performance, fabrication of electrode material is of significance. As Metal-organic frameworks (MOFs) imparts properties like high-tunability, favorable porous properties, tunable chemical compositions, controllable crystal structures, and adjustable geometry morphologies, which are favourable for the enhanced performance of SCs. The chapter provides a database of knowledge on how MOF with diversified physical and chemical features and functionalities can be utilized for the possible application as electrode material for supercapacitors. The chapter appreciates the choice of pristine MOF, MOF composites and MOF derived materials for an important application in the area of energy. Added to the above, the overview would give us scope for understanding and improving the properties of the MOF and hence the performance of the energy systems. Furthermore, it will kindle the interest to explore few more cost effective MOF for the energy storage application.

Keywords

Metal-Organic Framework, Electrode Material, Supercapacitor, Carbon Derived from MOF, Pristine MOF

Contents

Prospects of metal organic framework (MOF) as excellent electrode material for supercapacitors **94**

1.   Introduction..**95**

2.   Pristine MOF as electrode material **96**

    2.1   MOF with single metal and single ligand..96

    2.2   MOF with single metal and multiple ligands 97

2.3   MOF with multiple metals and single ligand 97

2.4   MOF with multiple metals and multiple ligands 97

3.   MOF Composites as Electrode Material for Supercapacitor **98**

4.   MOF-Derived Functional Materials as Supercapacitor Electrodes......................**99**

4.1   MOF as source for carbonaceous material 100

4.2   MOF as source for metal (oxide and hydroxide) particles 101

4.3   Metal sulfides and their composites 102

4.4   Metal selenides and their composites 103

4.5   Metal phosphides, metal carbides, metal nitrides and their composites............103

        xrGO/Ni$_2$P 104

Conclusion and future perspectives **104**

References...**105**

1.   Introduction

Due to population growth and industrial development, energy consumption by mankind is drastically increased. Depending more on the fossil fuels lead to its extinction and environmental damage. To meet today's energy demand, we move towards the use of clean, natural and renewable sources such as geothermal, solar and wind. Inconsistent power output [1] offered by these source limits its utility. Due to this, there is dire need for clean energy storage and conversion devices. So, migration towards electrochemical energy happens, which is highly reliable and efficient for practical applications. Recent research focuses mainly on the development of electrochemical energy devices and to improve its performance. Supercapacitors (SC), lithium – sulphur batteries (LSB), lithium - air batteries (LAB), sodium ion batteries (SIB) and lithium ion batteries (LIB) are the promising future generation electrochemical energy devices [2]. Among these devices SCs excels in the field of energy due to its enormous power density, long usage life and less re-energizing time [3-5]. Moreover its good rate performance, fast charge and discharge, energy saving, environmental protection, and low maintenance cost improves its applicability in electronic gadgets, motor vehicle, national defense and communication devices [6].

Short description about the working of SC is noteworthy here. Supercapacitors have two electrodes on either side of two compartments (double layer) of an electrolyte. In SCs both positive and negative electrodes are made up of the same material. Based on interfacial physics and chemistry, SCs are of three types namely electrical double-layer capacitors (EDLCs), pseudocapacitors and hybrid capacitors (HCs). EDLCs store charge through static electrical double layer formation (Helmholtz double-layer) with a high diffusion rate, thereby possesses high power density and excellent cyclability, but its applicability is limited due to its lower energy densities [7]. Higher capacitive capability of pseudocapacitors is due to the reversible Faradic redox reactions [8], but has low power density and cyclic stability [9]. HCs, formed by combination of EDLCs and pseudocapacitors, possess higher energy density and longer life span. The charge storage mechanism in SC is predominantly due to double layer charging effects. It is known that electrode material is one of the dominating factors influencing the performance of a supercapacitor. Electrode materials used in the capacitors [10] are of three main categories: carbon-based, transition metal oxides, and conducting polymers. Commonly, carbon-based active materials like CNTs, activated carbon (AC), graphene, and carbon black with large surface areas are suitable for EDLCs [11], whereas conducting polymers [12], transition metal oxides [13], and hydroxides [9] having good redox activity, various doping states, and high operating voltage are used as electrode materials for pseudocapacitors. Thus, it is essential to explore more novel electrode materials which can enhance energy density of SCs for the possible applications in the hybrid electric vehicles and portable energy systems.

Recently, Metal–organic frameworks (MOFs), MOFs are coordination polymers constructed with metal centers and organic ligands. These materials have gained attention due to its accessible high surface area, tunable pore sizes, open metal sites and ordered crystalline structures [14]. They have its utility in the field of such as magnetism [15], fluorescence [16] and catalysis [17]. In addition to that, MOFs are electrochemically active materials [18] thus find its utility in electrochemical fields including fuel cells [19], lithium-ion rechargeable batteries [20], supercapacitors [21], solar cells [22], Li–S batteries [23]. Recent research infers us, MOFs excels as excellent electrode materials for supercapacitors as MOFs has ultra-large surface areas ($\sim 10,000$ m^{2}g^{-1}), tunable pore sizes, various structures, and diverse functionalities [24, 25]. Moreover, MOFs are a type of bifunctional materials, which can be directly applied as both electrode active materials and templates to synthesize various nanostructures for electrochemical applications. Favorable products with desired structures could be generated through tuning the synthesis methods of MOFs.

This chapter gives an overview of the superior qualities of MOF and its possible application as electrode material for supercapacitor. Moreover, the utility of pristine MOF, MOF composites and MOF derived materials in SCs are presented in a systematic way.

2.   Pristine MOF as electrode material

The electrochemical performance of pristine MOFs is significantly dependent on their architectures and compositions. The architecture, including the frameworks and porous structure, affects the specific surface area, transport channel, electrical conductivity, and structure stability of MOFs, which is strongly related to their transport capability of electrons and ions. And the composition (especially the metal center) determines the electrochemical activity and the number of active sites, demonstrating the pseudocapacitive behavior of MOFs.

2.1   MOF with single metal and single ligand

Yaghi and Li in 1995 [26] introduced MOF, but as on date more than ten thousand MOFs are developed. On using MOF as electrode material for SCs, its porous nature, chemical structure and morphology are optimized by abundant organic ligands and reaction conditions. Metal ions/clusters and organic ligands in MOF provide redox sites for electrochemical reactions [27]. Though MOF possesses many qualities which favours electrochemical performance, its poor electronic conductivity imparts difficulty in direct usage of pristine MOFs as electrode materials for SCs.

Conductivity of MOF can be enhanced by constructing 3D network structures and conjugate systems, so that the well-designed pristine MOFs can be directly used as electrode materials in SCs. In earlier days, Co-MOFs [28] (Co-central metal atom and terephthalic acid as ligand) was found to be promising electrode materials for SCs as Co exists as multivalent metal ion. Moreover this MOF is of mesoporous nature which facilitates the ion diffusion in charge/discharge process. Further research continues with the development of various MOF by varying metal atoms and ligands. Electrochemical performance of these different MOFs depends on the electrolyte used. Further research was carried out by varying electrolytes.

2.2   MOF with single metal and multiple ligands

Later on new MOF was developed with single metal and multiple ligands to achieve enhanced performance. This type of MOF has complex microstructures with variable topological network and different pore size [35, 36] depending on the nature of ligand and the mode of coordination between metal and ligand. Synthesis of this type of MOF is laborious and many factors need to be controlled.

2.3   MOF with multiple metals and single ligand

Next stage is the development of MOF with multiple metals and single ligand, this type of material when used as electrode for SC produce higher specific capacitance as multi-metal ions contribute to redox charge storage and synergistic effect among components when the redox reaction occurs [41]. More valence changes in central metal ion and their mode of coordination offers different configurations of MOFs at different conditions [42]. Moreover, central metal ions have more valence changes and varied coordination modes, providing various configurations of MOFs at different conditions. As this type of MOF excels in its performance due to its better conductivity, better stability, or higher energy density [43-45], variety of multi-metal MOFs were synthesized.

2.4   MOF with multiple metals and multiple ligands

As a progression to improve the electrochemical performance, MOFs are synthesized by the combination of hetero metal and mixed ligand. As this type of MOFs has multiple redox reaction centers (due to the availability of multivalent metal) and a bridge for charge transfer. Moreover the synergistic effects between components also accounts for the enhanced electrochemical performance. Table 1 lists the electrochemical performance of pristine MOFs materials utilized as SCs electrode materials.

Table 1 Pristine MOFs as SCs electrode materials.

Material	Specific Capacitance (Fg^{-1})	Electrolyte	Potential window (V)	Capacitance retention (%)/Cycle number	Refs.
Pristine MOFs with single metal and single ligand					
Co-MOFs	206.76	LiOH	0.5	98.5/1000	[28]
Co(bpdc)(H$_2$O)$_2$·H$_2$O	179.2	LiOH	0.6	77.4/1000	[29]
Ni$_3$(btc)$_2$·12H$_2$O	726	KOH	0.4	94.6/1000	[30]
[Ni$_3$(OH)$_2$(C$_8$H$_4$O$_4$)2(H$_2$O)4]·2H$_2$O	988	KOH	0.45	96.5/5000	[31]
[Ni$_3$(OH)$_2$(C$_8$H$_4$O$_4$) 2(H$_2$O)$_4$]·2H$_2$O	1127	KOH	0.35	>90/3000	[32]
Cu-DBC	479	NaCl	0.7	72/1000	[33]
Cu-CAT	202	KCl	0.5	80/5000	[34]
Pristine MOFs with single metal and multiple ligands					
{[Co(Hmt)(tfbdc)(H$_2$O)$_2$]·(H$_2$O)$_2$}n	2474	KOH	0.5	94.3/2000	[37]
Ni$_2$(TATB)$_2$(L)$_2$(H$_2$O)	705	KOH	0.55	92.1/5000	[38]
[Co(2-ATA)2(4-bpdb)$_4$]n	325	KOH	0.4	92/6000	[39]
Ni(HBTC)(4,4'-bipy)	977	KOH	0.5	92.3/5000	[40]
Pristine MOFs with multiple metals and single ligand					
[Ni$_3$(OH)$_2$(C$_8$H$_4$O$_4$)$_2$ (H$_2$O)$_4$]·2H$_2$O	1620	KOH	0.35	91/3000	[43]
Co/Ni-MOF	236	KOH	0.6	-	[46]
NiCo-MOF	1126.7	KOH	0.5	93/3000	[47]
Co–Mn MOF	106.7	KOH	0.8	995/1500	[48]
Pristine MOFs with multiple metals and multiple ligands					
Ni/Co-MOF	758	KOH	0.48	75/5000	[49]
2D Ni/Co-MOF	2422.7	KOH	0.4	86/3000	[50]
Ni$_n$Co$_m$MOF	1333	KOH	0.5	79/2000	[51]
[Na$_2$Co(SDCA)(μ$_2$-OH)$_2$(μ$_2$-H$_2$O)$_2$(Azopy)]n	321.8	Na$_2$SO$_4$	1	97.4/5000	[52]
[NaZn$_2$(μ$_2$-BTC)$_2$ (μ$_2$-O)$_2$ (Azopy)(H$_2$O)$_2$]n	435.2	Na$_2$SO$_4$	2	100/4000	[53]
[(Ag$_7$bpy$_7$Cl$_2$) {AsW$_2$^{V}W$_{10}$^{VI}O$_{40}$}]·H$_2$O	986.1	H$_2$SO$_4$	1.05	1.4/5000	[54]

3.   MOF Composites as Electrode Material for Supercapacitor

Poor conductivity hampers the electrochemical performance of many MOFs, which can be overcome by MOF composites, i.e., by hybridization with other functional materials with good electrical conductivity, such as carbon materials, conducting polymers, and metals. The MOF based composites combine the advantages of the porous character of MOFs together with the functional characteristics of other components and may improve the electrochemical performance of pristine MOFs.

To prepare MOF- carbon composites, metal ions, ligands, and solvents are mixed with the carbon materials, and then the MOF will in situ grow on the surface of the carbon substrate for specific time and temperature [55]. Electrodeposition route is usually used to integrate MOFs with carbon materials and Ni, which can easily realize the in situ growth of MOFs uniformly on the substrate surface [56].

Emerging Materials for Next Frontier Energy and Environment Applications   Materials Research Forum LLC
Materials Research Foundations 170 (2024) 94-116   https://doi.org/10.21741/9781644903292-5

MOF – conducting polymer composites are divided into two categories. One is to grow MOFs on conducting polymer substrates, and the other is to grow conducting polymers on or into MOFs. The latter can be accomplished in a system containing a polymericmonomer, MOFs, an oxidant, and a solvent through chemical or electrochemical synthesis.

Many methods have been developed for the preparation of metal oxides/MOF composites. Chemically in situ self-transformation route was demonstrated to fabricate $MnO_x@MOF$ composite using MOF–Mn hexacyanoferrate hydrate nanocubes as the starting precursor [57].

Above said materials are binary composites, MOF-based ternary composites (composite made of three materials) can also be prepared via a similar procedure. For example, a two-step hydrothermal approach was performed to grow Ni-based MOFs on carbon cloth with Co_3O_4 [58]. A ternary graphene oxide (GO)/Cu-MOF/poly(3,4-ethylenedioxythiophene) (PEDOT) composite prepared via an electrochemical polymerization process [59]. An adhesive agent is added for the integration of MOFs with other functional materials and dopamine is the most widely used adhesive agent. List of MOF-Composites and its electrochemical performance in SCs are given in table 2.

Table 2 MOF-composites as SCs electrode materials

Material	Specific Capacitance (Fg^{-1})	Electrolyte	Potential window (V)	Capacitance retention (%)/Cycle number	References
Cu-MOF@δ-MnO_2	667	Na_2SO_4	1.03	96.3/6000	[60]
ZnO@MOF@PANI	340.7	KCl	1.1	82.5/5000	[61]
Co_3O_4@Co-MOF	1020	KOH	0.7	96.7/5000	[62]
Ni-MOF@$Co(OH)_2$	1448	KOH	1.4	87.3/8000	[63]
PANI-ZIF-67-CC	371	KCl	1.2	80/2000	[64]
CNTs@Mn-MOF	203.1	Na_2SO_4	1.0	88/3000	[65]
Ni-MOF/CNT	1765	KOH	1.6	95/5000	[66]
ZIF–PPy	554	Na_2SO_4	0.6	>90/10000	[67]
rGO/HKUST-1	385	$NaNO_3$	0.8	98.5/4000	[68]
GA@UiO-66-NH_2	651	Na_2SO_4	1.1	88/10000	[69]
Co-doped Cu-MOF/$Cu_{2+1}O$	518.58	KOH	0.5	97.26/5000	[70]
rGO–Ni-doped MOF	758	KOH	0.5	21.4/500	[71]
PPNF@Co–Ni MOF	1096.2	KOH	0.5	-	[72]
Zn/Ni-MOF@PPy	160.1	KOH	0.36	94.4/3000	[73]

4. MOF-Derived Functional Materials as Supercapacitor Electrodes

MOFs have great potential to convert into other types of active electrode materials, which can be due to the following reasons. On one hand, the metal center and organic ligand can provide metal, carbon, metal oxides, metal hydroxides, metal sulfides, metal phosphates, metal selenides, metal carbides, etc. as active components for supercapacitors. Therein, nitrogen, sulfur, and phosphorus sources have also been used as dopants in carbon materials. On the other hand, a part of the framework and pore structure of MOFs can be preserved after a series of treatments, resulting in abundant exposure of electrochemical active sites which favors the ions transportation.

4.1   MOF as source for carbonaceous material

Carbon materials derived from MOFs show superiority to conventional carbon materials because of their distinct compositions and unique porous structures. Typically, organic ligands in MOFs as precursors not only offer a rich carbon source, but also grant carbon materials with various heteroatom doping, contributing to a high specific capacitance, which is related to the pseudocapacitive effect caused by N, P, and O dopants.

In recent years, carbonaceous materials obtained by direct carbonization of organic materials, have been used in SCs due to their nontoxic nature and excellent electrical conductivity, but its applications are limited due to the low surface areas, non-uniform particle sizes, and chaotic structures [74]. These can be overlooked by using MOF-derived carbon nanocomposites with nanoporous architectures and ultra-large ionic accessible surface areas [75]. As the micro- and mesoporous carbonaceous materials like graphene, AC, and CNTs have large interspace which store enough electrolyte and shorten the diffusion path for ion transportation thereby plays a significant role in SCs [76]. In addition to that, in situ heteroatoms doping (such as P, F B, N, and S) in prepared carbonaceous nanocomposites is possible by suitable selection of MOF precursors. Heteroatom doping enhances the electrical conductivity, cyclability, and capacitive performance of nanoporous carbons (NPCs) due to the synergistic effect [77-80]. As a result of it MOF-derived carbons have its widespread application as electrode material in SCs.

Table 3 MOF-derived carbon based electrodes

Material	Specific Capacitance (Fg^{-1})	Electrolyte	Potential window (V)	Capacitance retention (%)/Cycle number	References
NPC	204	H_2SO_4	1	95/6000	[81]
NPC350	158	H_2SO_4	1	-	[82]
MC-A	208	KOH	1	-	[83]
MAC-A	271	KOH	1	-	
C-MOF-2	170	H_2SO_4	1	-	[84]
C800	188	H_2SO_4	1	-	[85]
C1000	161	H_2SO_4	1	-	
Z-900	257	H_2SO_4	1.2	99/250	[86]
AS-ZC-800	211	H_2SO_4	1	>90/2000	[87]
NPC	226	H_2SO_4	0.9	92/2000	[88]
NC@GC	216	H_2SO_4	0.8	99/10000	[89]
MOF-DC	238	$LiPF_6/EC/DMC$	1	82/10000	[90]
NPC	425	H_2SO_4	1	-	[91]
NC900	367	KOH	1	95/3000	[92]
WNPC	300.7	H_2SO_4	1	87/10000	[93]
UCNs	256	KOH	1	91/10000	[94]
Hybrid-porous C	215	6M KOH	1	100/10000	[95]
N-Doped C	244	1 M H_2SO4	1	98.5/10000	[96]
PC_Zn	138	6M KOH	1	91.8/5000	[97]
c-CN	333.8	1 M LiTFSI	1.2	88.1/12000	[98]

NPCs (Nano Porous Carbon) can be obtained from MOFs through a simple heat treatment, and these active materials often exhibit micro/ mesoporous architectures and large specific surface areas, which are crucial characteristics to optimize the electrochemical performance for SCs. Thus, the as-derived carbon materials could have good electrochemical performance. The electrochemical property of MOF-derived carbon can be further improved by the addition of heteroatoms that are derived from organic ligands. These heteroatoms are able to strengthen the interactions between ions and surfaces, which is beneficial to stabilize the porous structures. The high capacitance could be caused by the pseudo-capacitance of functional groups and unique structure. Firstly, oxygen-containing groups, forming during the process of pyrolysis, were not only beneficial to generate pseudo-capacitance, but also enhanced the wettability of active materials. Nitrogen-containing functional groups could be responsible for the redox pseudo-capacitance due to the existence of pyridinic and pyrrolic groups. Secondly, the hierarchical micro- and meso-porous architecture with a large number of sp^2 bonded carbon accelerated the diffusion of ions and eventually enhanced the capacitive performance. Lastly, it provided sufficient electroactive sites for charge transport and interactions. Electrochemical performance of MOF-derived carbon material as electrode active material for SCs is listed in table 3.

4.2   MOF as source for metal (oxide and hydroxide) particles

In the molecular structure of MOFs, metal centers are in coordination with organic ligands. The MOFs-derived metal compounds with unique structures can be prepared by controlled calcination under certain environments. Generally, metal compounds as pseudocapacitors have better energy density and electron capacity than conventional porous carbon materials. Ideal metal compound electrodes usually possess several requirements, such as good electrical conductivity, high porosity, large surface area, and robust chemical and thermal stability. Fortunately, a large number of metal compounds originating from the MOFs precursors or templates meet these conditions.

Compared with the metal oxides prepared by other methods, MOF-derived metal oxides exhibit some unique merits. First, the diversity of composition and morphology of metal oxides derived from MOF precursors enables metal oxides to achieve more possibilities as high-performance electrode materials in SCs. Second, the high surface areas and novel architecture of MOF-derived metal oxides provide an efficient diffusion path for electron and ion transfer. As a pseudocapacitive material, metal oxides possess abundant redox active sites, whose amount and activity make an important impact on their specific capacitances. Furthermore, the design of MOF-derived metal oxides with novel architectures and compositions is also suggested to improve the electrochemical performance.

A large number of transition metal oxides (Co_3O_4, FeO_x, MnO_x, NiO, CeO_2, CuO_x, and ZnO) can be derived from different kinds of MOFs as electrode materials for supercapacitors. Their redox activities are related to metal species and valence states. Among these suitable metal oxides, manganese oxides, involving various oxidation states, are one the most promising electrode materials for supercapacitors due to their fast redox charge transfer, which can be realized from MOFs under different calcination temperatures. The different calcination process can also achieve diversity of chemical composition and morphology of the MOF-derived MnO_x.

Transition metal hydroxides show similar electrochemical features to metal oxides, such as large theoretical capacity, low electrical conductivity, and modest reaction kinetics. The phase transformation process from MOFs to metal hydroxides plays a key role in their electrochemical performance; thus, the related parameters, such as the etching time, temperature, and concentration of the alkaline solution, have been extensively studied. The transformation process is usually accompanied by the formation of a hollow structure, which stimulates electron/ion transfer.

Table 4 MOF-derived metal oxides and hydroxides based electrodes

Material	Specific Capacitance (Fg^{-1})	Electrolyte	Potential window (V)	Capacitance retention (%)/Cycle number	References
Porous Co_3O_4 polyhedron	110	KOH	0.5	83.15/20	[99]
Co_3O_4	233	KOH	0.5	89.8/1300	[100]
Porous Co_3O_4	148	KOH	0.5	68.8/1500	[101]
Co_3O_4	706	KOH	0.4	71.5/1000	[102]
Co_3O_4	647	KOH	0.5	99/1500	[103]
Co_3O_4	900	KOH	0.7	85.5/20000	[104]
Porous $Co(OH)_2$	604.5	KOH	0.6	70/2000	[105]
$Co(OH)_2$	688	KOH	1.5	94/5000	[106]
NiO nanospheres	473	KOH	1.5	94/3000	[107]
MnO_x	220	Na_2SO_4	1.0	94.37/2000	[108]
MoO_3/CuO	251.8	LiOH	0.8	88.5/5000	[109]
WO_3/CuO	248.2	KOH	0.5	85.2/1500	[110]
CeO2	1205	KOH	0.5	~100/5000	[111]
Fe3O4	139	KOH	1	83.3/4000	[112]
Cr_2O_3	180	KOH	1	95.5/3000	[113]
$Cu–Cu_2O–CuO$	750	KOH	0.6	94.5/3000	[114]
$Co(VO_3)_2$ - $Co(OH)2$	803	KOH	0.6	90/15 000	[115]
$NiCo_2O_4$	1055.3	KOH	1.5	86.7/20000	[116]
$MOF-Co_9S_8$	2504	KOH	0.5	95.3/ 8000	[117]
Ni–Zn–Co–S nano sword arrays	1289	KOH	0.7	85/1000	[118]
$Ni_xCo_{3-x}O_4$	2870.8	KOH	1.5	81/ 5000	[119]

4.3   Metal sulfides and their composites

Transition metal sulfides are considered promising for SC electrode materials due to their higher electrochemical activity and superior electrical conductivity than other metal oxides or hydroxides [120,121]. Because of the lower electronegativity of sulfur compared with that of oxygen, sulfides tend to have more flexible structures and enhanced electron transfer compared with those of oxides [122]. According to the synthetic routes, metal sulfides derived from MOFs can be classified into two categories. One is obtained through a solvothermal reaction. The inherent characteristics of the original MOFs, including the morphology, high surface area, and porous structure, can be preserved after their conversion to metal sulfides. The other is generated through a calcination process, corresponding to a metal sulfide/carbon composite with better electrical conductivity. Over the last few years, MOF-derived metal sulfides and their composites with controllable compositions and specific architecture have been proven to show excellent electrochemical performance for supercapacitor application.

Table 5 MOF-derived metal sulfides based electrodes

Material	Specific Capacitance (Fg^{-1})	Electrolyte	Potential window (V)	Capacitance retention (%)/Cycle number	References
CoS	980	KOH	1.6	88/10000	[123]
CoS$_{1.097}$/N-doped carbon	360.1	KOH	0.6	~90/2000	[124]
Co$_9$S$_8$@S,N-doped carbon	429	KOH	0.8	98/2000	[125]
Co$_9$S$_8$/carbon	734.1	KOH	0.5	99.8/ 140 000	[126]
Ni$_3$S$_2$ nanosheet	1000	NaOH	1.8	62/2000	[127]
NiS/rGO	1488	KOH	0.6	89/20000	[128]
NiS$_2$/ZnS nanospheres	1198	KOH	0.6	87/1000	[129]
NiS nanocube	2112	KOH	0.5	91.8/4000	[130]
NiCo$_2$S$_4$	1382	KOH	0.5	79/ 10,000	[131]
Ni/Ni$_3$S$_2$/CNF	830	PVA–KOH	0.8	95.7/ 5000	[132]

4.4   Metal selenides and their composites

Metal selenides exhibit higher conductivity, lower electronegativity, faster reaction kinetics, and higher volume-specific capacity than their sulfide counterparts [133]. These characteristics make metal selenides attractive candidates in the field of supercapacitors. Many researches infer that the electrochemical performance of single metallic selenide is not superior over bimetallic selenides due to the synergistic effects between bimetallic metal selenides, and the electronic structure and coupling interaction can be more probably adjusted. Therefore, bimetallic selenides are promising electrode materials for supercapacitors. Nevertheless, the low conductivity and limited exposure of active sites make their electrochemical activity not efficient to display. The increment of utilization on the redox sites of two metal selenides through rational design of the structure is therefore of great significance towards promoted electrochemical performances.

Table 6 MOF-derived metal selenides based electrodes

Material	Specific capacity mAh g^{-1}/Specific Capacitance (Fg^{-1})	Electrolyte	Potential window (V)	Capacitance retention (%)/Cycle number	References
H-Ni-Co-Se	175 Fg^{-1}	-	0.8	89.3/2000	[134]
(Ni$_{0.33}$Co$_{0.67}$)Se$_2$	827.9 Fg^{-1}	KOH	0.6	113.6/2000	[135]
Co–Mo–Se	221.7mAh g^{-1}	KOH	0.6	95/8000	[136]
MNSe@NF	325.6 mA h g^{-1}	KOH	0.7	96.8/15000	[137]
Zn–Ni–Se/Ni(OH)$_2$	1632.8 Fg^{-1}	KOH	0.6	85.4/2000	[138]
CoSe$_2$/NC-400	120.2 mA h g^{-1}	KOH	0.4	92/10000	[139]

4.5   Metal phosphides, metal carbides, metal nitrides and their composites

Along with the rapid development of MOF-derived materials for SCs, a large variety of other MOF-derived materials, such as metal phosphides, metal nitrides, and metal carbides, have stimulated significant consideration of researchers. In general, the lower electronegativity of P than O in their respective compounds is conductive to electron transport and redox reactions, resulting in the ideal electrical conductivity and high theoretical capacitance of the transition metal phosphide. Moreover, the wide band gap and poor electrical conductivity of the metal oxides and metal hydroxides lead to poor electrochemical activity and rate performance limits their application and development. Recently, transition-metal phosphides have become the most significant alternative materials in the diversified fields of overall water splitting and SCs. Compared with monometallic phosphides, bimetallic phosphides exhibit a higher specific capacitance because of their rich valence states and fast charge transfer rate.

Metal nitrides are widely recognized as a class of desirable supercapacitor electrode materials owing to their high electrical conductivity and structural stability. Embedding metal nanoparticles in nitrides can further enhance the conductivity of electron transport.

In addition, Ni-doped Co–Co_2N and sheet-like Cr_3C_2 derived from the corresponding

MOFs have also been investigated as electrode materials, showing good electrochemical behavior [142].

Table 7 MOF-derived metal phosphide/nitride/carbide based electrodes

Material	Specific Capacitance (Fg^{-1})	Electrolyte	Potential window (V)	Capacitance retention (%)/Cycle number	References
$Zn_{0.33}Co_{0.67}P$	1086.5	KOH	0.55	80.3/7000	[140]
xrGO/Ni_2P	890	KOH	0.6	66.7/8000	[141]
Ni-doped Co–Co_2N	361.93	KOH	0.6	82.4/5000	[142]

Conclusion and future perspectives

This chapter gives an overview of the performance exhibited by MOF materials, their composites and its derivatives as electrode material of SCs. Though these materials have favourable qualities like large surface areas, high porosity, tunable sizes and structures, there are still some persistent challenges that need to be tackled comprehensively. The electrical conductivity of pristine MOFs can be enhanced by incorporating highly conductive nanocomposites due to synergistic effects. Secondly, most MOF nanocomposites cannot possess the properties of high energy density and power density simultaneously; hence crucial for researchers to systematically explore advanced characterization tools and electrochemical mechanisms. Additionally, studies on the electrochemical performance of MOFs with multi-metal ions and multi-organic ligands are still rare, and great efforts should be made on this aspect. Then, electrolytes have a significant role to play in the electrochemical performance. The electrochemical performances of most MOF nanomaterial are studied in aqueous and organic-based electrolytes and ionic electrolytes with distinctive electrochemical stability, and wide operating potential windows might able to offer realization on the development of a safe, durable, and sustainable SC system.

The widespread application of MOFs derived electrode materials for commercial or industrial supercapacitor application is retarded by the slow synthesis process and high cost of the synthesis of MOFs precursors and MOFs-derived materials. Thus, it is important to develop new synthetic routes in the MOF production with low cost and high scalability, the synthesis operation and process should be efficient, relatively fast, and low labor cost.

Furthermore, the integration of these MOF-derived materials with other functional materials should be given considerable attention to realize synergistically enhanced electrochemical performance. Moreover, MOF based materials as flexible electrodes often depend on other flexible substrates. It is of great importance to achieve a balance between electrochemical performance and mechanical properties. In addition, exploring a simple and versatile technology to achieve an efficient interaction between the two materials is also a promising direction, which is conducive to promoting the early realization of industrial production of these materials.

References

[1] L. Xiao, L. Lin, Y. Liu, Discussions on the architecture and operation mode of future power grids, Energies 4 (2011) 1025-1035. https://doi.org/10.3390/en4071025

[2] H. Zhang, A. Chen, M. Zhong, Z. Zhang, X. Zhang, Z. Zhou, X.H. Bu, Metal-organic frameworks (MOFs) and MOF-derived materials for energy storage and conversion, Electrochem. Energy Rev. 2 (2019) 29−104. https://doi.org/10.1007/s41918-018-0024-x

[3] B.K.Kim, S. Sy, A. Yu, J. Zhang, Electrochemical Supercapacitors for energy storage and conversion, Handbook Clean. Energ. Syst. 5 (2015) 1-25. https://doi.org/10.1002/9781118991978.hces112

[4] Y.Wang, Y. Song, Y. Xia, Electrochemical capacitors: Mechanism, materials, systems, characterization and applications. Chem. Soc. Rev. 45, (2016) 5925-5950. https://doi.org/10.1039/C5CS00580A

[5] G.Gautham Prasad, N. Nidheeshshetty, S. Thakur, Rakshitha, K.B. Bommegowda, Supercapacitor Technology and its Applications: A Review. 2019 IOP Conf. Ser.: Mater. Sci. Eng. 561, 012105 https://doi.org/10.1088/1757-899X/561/1/012105

[6] L. G. H. Staaf, P. Lundgren, P. Enoksson, Present and future supercapacitor carbon electrode materials for improved energy storage used in intelligent wireless sensor systems, Nano Energy 9 (2014) 128−141. https://doi.org/10.1016/j.nanoen.2014.06.028

[7] L. Zeng, X. Lou, J. Zhang, C. Wu, J. Liu, C. Jia, Carbonaceous mudstone and lignin-derived activated carbon and its application for supercapacitor electrode. Surf. Coat. Technology 357 (2019) 580-586. https://doi.org/10.1016/j.surfcoat.2018.10.041

[8] Y. Jiang, J. Liu, Definitions of pseudocapacitive Materials: A brief review, Energy Environ. Mater. 2 (2019) 30-37. https://doi.org/10.1002/eem2.12028

[9] R.R. Salunkhe, Y.V. Kaneti, Y. Yamauchi, Metal-Organic Framework-derived nanoporous metal oxides toward supercapacitor applications: Progress and prospects, ACS Nano. 11 (2017) 5293-5308. https://doi.org/10.1021/acsnano.7b02796

[10] Z. Yong, F. Hui, W. Xingbing, W. Lizhen, Progress of electrochemical capacitor electrode materials: a review, Int. J. Hydrog. Energy 34 (2009) 4889- 4899. https://doi.org/10.1016/j.ijhydene.2009.04.005

[11] R.B. Rakhi, W. Chen, D. Cha, H.N. Alshareef, Nanostructured ternary electrodes for energy-storage applications, Adv. Energ. Mater. 2 (2012) 381-389. https://doi.org/10.1002/aenm.201100609

[12] M. Boota, Y. Gogotsi, MXene-Conducting polymer asymmetric pseudocapacitors, Adv. Energ. Mater. 9 (2019) 1802917. https://doi.org/10.1002/aenm.201802917

[13] S. Zheng, X. Li, B. Yan, Q. Hu, Y. Xu, X. Xiao, H. Xue, H. Pang, Transition-Metal (Fe, Co, Ni) based metal-organic frameworks for electrochemical energy storage, Adv. Energ. Mater. 7 (2017) 1602733. https://doi.org/10.1002/aenm.201602733

[14] S. L. James, Metal-organic frameworks, Chem. Soc. Rev. 32 (2003)276-288. https://doi.org/10.1039/b200393g

[15] M. Kurmoo, Magnetic metal-organic frameworks, Chem. Soc. Rev. 38 (2009) 1353-1379. https://doi.org/10.1039/b804757j

[16 L. Chen, K. Tan, Y.Q. Lan, S.L. Li, K.Z. Shao Z.-M. Su, Unusual microporous polycatenane-like metal-organic frameworks for the luminescent sensing of Ln3+cations and rapid adsorption of iodine, Chem. Commun. 48 (2012) 5919-5921. https://doi.org/10.1039/c2cc31257c

[17] R.Q. Zou, H. Sakurai, S. Han, R.Q. Zhong, Q. Xu, Probing the Lewis Acid Sites and CO Catalytic Oxidation Activity of the Porous Metal−Organic Polymer [Cu(5-methylisophthalate)], J. Am. Chem. Soc. 129 (2007) 8402-8403. https://doi.org/10.1021/ja071662s

[18] A. Morozan and F. Jaouen, Metal organic frameworks for electrochemical applications, Energy Environ. Sci. 5 (2012) 9269-9290. https://doi.org/10.1039/c2ee22989g

[19] X. Q. Liang, F. Zhang, W. Feng, X. Q. Zou, C. J. Zhao, H. Na, C. Liu, F. X. Sun, G. S. Zhu, From metal-organic framework (MOF) to MOF-polymer composite membrane: enhancement of low-humidity proton conductivity, Chem. Sci. 4, (2013) 983-992. https://doi.org/10.1039/C2SC21927A

[20] A. Banerjee, K. K. Upadhyay, D. Puthusseri, V. Aravindan, S. Madhavi, S. Ogale, MOF-derived crumpled-sheet-assembled perforated carbon cuboids as highly effective cathode active materials for ultra-high energy density Li-ion hybrid electrochemical capacitors (Li-HECs), Nanoscale 6 (2014) 4387-4394. https://doi.org/10.1039/c4nr00025k

[21] S. Dutta, A. Bhaumik and K. C.-W. Wu, Hierarchically porous carbon derived from polymers and biomass: effect of interconnected pores on energy applications, Energy Environ. Sci. 7 (2014) 3574-3592 https://doi.org/10.1039/C4EE01075B

[22] A. V. Vinogradov, H. Z. Hertling, E. H. Hawkins, A. V. Agafonov, G. A. Seisenbaeva, V. G. Kessler, V. V. Vinogradov, The first depleted heterojunction TiO2-MOF-based solar cell, Chem. Commun., 50 (2014) 10210-10213. https://doi.org/10.1039/C4CC01978D

[23] J. W. Zhou, R. Li, X. X. Fan, Y. F. Chen, R. D. Han, W. Li, J. Zheng, B. Wang and X. G. Li, Rational design of a metal-organic framework host for sulfur storage in fast, long-cycle Li-S batteries, Energy Environ. Sci. 7 (2014) 2715-2724. https://doi.org/10.1039/C4EE01382D

[24] P.M. Schoenecker, C.G. Carson, H. Jasuja, C.J.J. Flemming, K.S. Walton, Effect of water adsorption on retention of structure and surface area of metal-organic frameworks, Ind. Eng. Chem. Res. 51, (2012) 6513-6519. https://doi.org/10.1021/ie202325p

[25] I. Senkovska, S.Kaskel, Ultrahigh porosity in mesoporous MOFs: Promises and limitations, Chem. Commun. (Camb). 50, (2014) 7089-7098. https://doi.org/10.1039/c4cc00524d

[26] O.M. Yaghi, H. Li, Hydrothermal synthesis of a metal-organic framework containing large rectangular channels, J. Am. Chem. Soc. 117 (1995) 10401−10402. https://doi.org/10.1021/ja00146a033

[27] A.A. Talin, A. Centrone, A.C. Ford, M.E. Foster, V. Stavila, P.Haney, R.A. Kinney, V. Szalai, F. El Gabaly, H.P. Yoon, F. Leonard, M.D. Allendorf, Tunable electrical conductivity in metal organic framework thin-film devices, Science 343 (2014) 66−69. https://doi.org/10.1126/science.1246738

[28] D.Y. Lee, S.J. Yoon, N.K. Shrestha, S.H. Lee, H. Ahn, S.H Han, Unusual energy storage and charge retention in Co-based metal-organic-frameworks, Microporous Mesoporous Mater. 153, (2012) 163−165. https://doi.org/10.1016/j.micromeso.2011.12.040

[29] D.Y. Lee, D.V. Shinde, E.K. Kim, W. Lee, I.W. Oh, N.K. Shrestha, J.K. Lee, S.H. Han, Supercapacitive property of metal-organic-frameworks with different pore dimensions and Morphology, Microporous Mesoporous Mater. 171 (2013) 53−57. https://doi.org/10.1016/j.micromeso.2012.12.039

[30] L. Kang, S.X. Sun, L.B. Kong, J.W. Lang, Y.C. Luo, Investigating metal-organic framework as a new pseudo-capacitive material for supercapacitors, Chin. Chem. Lett. 25 (2014) 957−961. https://doi.org/10.1016/j.cclet.2014.05.032

[31] Y.Yan, P. Gu, S. Zheng, M. Zheng, H. Pang, H. Xue, Facile synthesis of an accordion-like Ni-MOF superstructure for high performance flexible supercapacitors, J. Mater. Chem. A 4 (2016) 19078−19085. https://doi.org/10.1039/C6TA08331E

[32] J. Yang, P. Xiong, C. Zheng, H. Qiu, M. Wei, Metal-organic frameworks: A new promising class of materials for a high performance supercapacitor electrode, J. Mater. Chem. A 2 (2014) 16640−16644. https://doi.org/10.1039/C4TA04140B

[33] J. Liu, Y. Zhou, Z. Xie, Y. Li, Y. Liu, J. Sun, Y. Ma, O. Terasaki, L. Chen, Conjugated copper-catecholate framework electrodes for efficient energy storage, Angew. Chem. Int. Ed. 59 (2020) 1081−1086. https://doi.org/10.1002/anie.201912642

[34] W.H. Li, K. Ding, H.R. Tian, M.S. Yao, B. Nath, W.H. Deng, Y. Wang, G. Xu, Conductive metal-organic framework nanowire array electrodes for high-performance solid-state supercapacitors, Adv. Funct. Mater. 27 (2017) 1702067. https://doi.org/10.1002/adfm.201702067

[35] J.P. Meng, Y. Gong, Q. Lin, M.M. Zhang, P. Zhang, H.F. Shi, J.H. Lin, Metal-organic frameworks based on rigid ligands as separator membranes in supercapacitor, Dalton Trans. 44 (2015) 5407−5416. https://doi.org/10.1039/C4DT03702B

[36] C. Qu, Y. Jiao, B. Zhao, D. Chen, R. Zou, K.S. Walton, M. Liu, Nickel-based pillared MOFs for high-performance supercapacitors: Design, synthesis and stability study, Nano Energy 26 (2016) 66−73. https://doi.org/10.1016/j.nanoen.2016.04.003

[37] X. Liu, C. Shi, C. Zhai, M. Cheng, Q. Liu, G. Wang, Cobalt based layered metal-organic framework as an ultrahigh capacity supercapacitor electrode material, ACS Appl. Mater. Interfaces 8 (2016) 4585−4591. https://doi.org/10.1021/acsami.5b10781

[38] K. Wang, Z. Wang, X. Wang, X. Zhou, Y. Tao, H. Wu, Flexible long-chain-linker constructed Ni-based metal-organic frameworks with 1D helical channel and their pseudo-

capacitor behavior studies, J. Power Sources 377 (2018) 44−51.
https://doi.org/10.1016/j.jpowsour.2017.11.087

[39] R. Abazari, S. Sanati, A. Morsali, A. Slawin, C.L. Carpenter-Warren, Dual-purpose 3D pillared metal-organic framework with excellent properties for catalysis of oxidative desulfurization and energy storage in asymmetric supercapacitor, ACS Appl. Mater. Interfaces 11 (2019), 14759−14773. https://doi.org/10.1021/acsami.9b00415

[40] Y. Li, Y. Xu, Y. Liu, H. Pang, Exposing {001} crystal plane on hexagonal Ni-MOF with surface-grown cross-linked mesh-structures for electrochemical energy storage, Small 15 (2019) 1902463. https://doi.org/10.1002/smll.201902463

[41] H. Yu, H. Xia, J. Zhang, J. He, S. Guo, Q. Xu, Fabrication of Fe-doped Co-MOF with mesoporous structure for the optimization of supercapacitor performances, Chin. Chem. Lett. 29 (2018) 834−836. https://doi.org/10.1016/j.cclet.2018.04.008

[42] F. Xu, N. Chen, Z. Fan, G. Du, Ni/Co-based metal organic frameworks rapidly synthesized in ambient environment for high energy and power hybrid supercapacitors, Appl. Surf. Sci. 528 (2020) 146920. https://doi.org/10.1016/j.apsusc.2020.146920

[43] J. Yang, C. Zheng, P. Xiong, Y. Li, M. Wei, Zn-doped Ni- MOF material with a high supercapacitive performance, J. Mater. Chem. A 2 (2014) 19005−19010. https://doi.org/10.1039/C4TA04346D

[44] H. Xia, J. Zhang, Z. Yang, S. Guo, S. Guo, Q. Xu, 2D MOF nanoflake-assembled spherical microstructures for enhanced supercapacitor and electrocatalysis performances, Nano-Micro Lett. 9 (2017) 43. https://doi.org/10.1007/s40820-017-0144-6

[45] C.Ye, Q. Qin, J. Liu, W. Mao, J. Yan, Y. Wang, J. Cui, Q. Zhang, L. Yang, Y. Wu, Coordination derived stable Ni-Co MOFs for foldable all-solid-state supercapacitors with high specific energy, J. Mater. Chem. A 7 (2019) 4998−5008 https://doi.org/10.1039/C8TA11948A

[46] Y.Jiao, J. Pei, D. Chen, C. Yan, Y. Hu, Q. Zhang, G. Chen, Mixed-metallic MOF based electrode materials for high performance hybrid supercapacitors, J. Mater. Chem. A 5 (2017) 1094−1102. https://doi.org/10.1039/C6TA09805C

[47] J. Sun, X. Yu, S. Zhao, H. Chen, K. Tao, L. Han, Solvent controlled morphology of amino-functionalized bimetal metal-organic frameworks for asymmetric supercapacitors, Inorg. Chem. 59 (2020) 11385−11395. https://doi.org/10.1021/acs.inorgchem.0c01157

[48] S. H. Kazemi, B. Hosseinzadeh, H. Kazemi, M.A. Kiani, S. Hajati, Facile synthesis of mixed metal-organic frameworks: Electrode materials for supercapacitors with excellent areal capacitance and operational stability, ACS Appl. Mater. Interfaces 10 (2018) 23063−23073. https://doi.org/10.1021/acsami.8b04502

[49] S. Gao, Y. Sui, F. Wei, J. Qi, Q. Meng, Y. Ren, Y. He, Dandelion-like nickel/cobalt metal-organic framework based electrode materials for high performance supercapacitors, J. Colloid Interface Sci. 531 (2018) 83−90. https://doi.org/10.1016/j.jcis.2018.07.044

[50] T. Sun, L. Yue, N. Wu, M. Xu, W. Yang, H. Guo, W. Yang, Isomorphism combined with intercalation methods to construct a hybrid electrode material for high-energy storage capacitors, J. Mater. Chem. A 7 (2019) 25120−25131. https://doi.org/10.1039/C9TA08696J

[51] Y. Liang, W. Yao, J. Duan, M. Chu, S. Sun, X. Li, Nickel cobalt bimetallic metal-organic frameworks with a layer-and-channel structure for high-performance supercapacitors, J.Energy Storage 33 (2021) 102149. https://doi.org/10.1016/j.est.2020.102149

[52] R.Rajak, M. Saraf, S.M. Mobin, Mixed-ligand architected unique topological heterometallic sodium/cobalt-based metal-organic framework for high-performance supercapacitors, Inorg. Chem. 59 (2020) 1642−1652. https://doi.org/10.1021/acs.inorgchem.9b02762

[53] R.Rajak, M. Saraf, S.M. Mobin, Robust heterostructures of a bimetallic sodium zinc metal-organic framework and reduced grapheme oxide for high-performance supercapacitors, J. Mater. Chem. A 7 (2019) 1725−1736. https://doi.org/10.1039/C8TA09528K

[54] L. Cui, K. Yu, J. Lv, C. Guo, B. Zhou, A 3D POMOF based on a {AsW12} cluster and a Ag-MOF with interpenetrating channels for large-capacity aqueous asymmetric supercapacitors and highly selective biosensors for the detection of hydrogen peroxide, J. Mater. Chem. A 8 (2020) 22918−22928. https://doi.org/10.1039/D0TA08759A

[55] J. Feng, L. Liu, Q. Meng, Enhanced electrochemical and capacitive deionization performance of metal organic framework/holey graphene composite electrodes, J. Colloid Interface Sci. 582, (2021) 447-458. https://doi.org/10.1016/j.jcis.2020.08.091

[56] S. C. Wechsler, F. Z. Amir, Superior electrochemical performance of pristine nickel hexaaminobenzene mof supercapacitors fabricated by electrophoretic deposition ChemSusChem 13, (2020) 1491-1495. https://doi.org/10.1002/cssc.201902691

[57] Y. Z. Zhang, T. Cheng, Y. Wang, W. Y. Lai, H. Pang, W. Huang, A simple approach to boost capacitance: flexible supercapacitors based on manganese oxides@mofs via chemically induced in situ self-transformation,Adv. Mater. 28 (2016) 5242-5248. https://doi.org/10.1002/adma.201600319

[58] L. Zhang, Y. Zhang, S. Huang, Y. Yuan, H. Li, Z. Jin, J. Wu, Q. Liao, L. Hu, J. Lu, S. Ruan, Y. J. Zeng, Co3O4/Ni-based MOFs on carbon cloth for flexible alkaline battery-supercapacitor hybrid devices and near-infrared photocatalytic hydrogen evolution, Electrochim. Acta 281, (2018) 189-197. https://doi.org/10.1016/j.electacta.2018.05.162

[59] D. Fu, H. Li, X. M. Zhang, G. Han, H. Zhou, Y. Chang, Flexible solid-state supercapacitor fabricated by metal-organic framework/graphene oxide hybrid interconnected with PEDOT, Mater. Chem. Phys. 179, (2016) 166-173. https://doi.org/10.1016/j.matchemphys.2016.05.024

[60] J. Xu, Y. Wang, S. Cao, J. Zhang, G. Zhang, H. Xue, Q. Xu, H. Pang, Ultrathin Cu-MOF@delta-MnO2 nanosheets for aqueous electrolyte-based high-voltage electrochemical capacitors. J. Mater.Chem. A 6 (2018) 17329−17336. https://doi.org/10.1039/C8TA05976D

[61] C. Zhu, Y. He, Y. Liu, N. Kazantseva, P. Saha, Q. Cheng, ZnO@MOF@PANI core-shell nanoarrays on carbon cloth for high performance supercapacitor electrodes. J. Energy Chem. 35(2019) 124−131. https://doi.org/10.1016/j.jechem.2018.11.006

[62] S. Zheng, Q. Li, H. Xue, H. Pang, Q. Xu, A highly alkaline stable metal oxide@metal-organic framework composite for high performance electrochemical energy storage. Natl. Sci. Rev. 7(2020), 305−314. https://doi.org/10.1093/nsr/nwz137

[63] X. Shi, T. Deng, G. Zhu, Vertically oriented Ni-MOF@ Co(OH)2 flakes towards enhanced hybrid supercapacitior performance, J. Colloid Interface Sci. 593, (2021) 214−221. https://doi.org/10.1016/j.jcis.2021.02.096

[64] L. Wang, X. Feng, L. Ren, Q. Piao, J. Zhong, Y. Wang, H. Li, Y. Chen, B. Wang, Flexible solid-state supercapacitor based on a metal-organic framework interwoven by electrochemically-deposited PANI, J. Am. Chem. Soc. 137 (2015) 4920−4923. https://doi.org/10.1021/jacs.5b01613

[65] Y. Zhang, B. Lin, Y. Sun, X. Zhang, H. Yang, J. Wang, Carbon nanotubes@metal-organic frameworks as Mn-based symmetrical supercapacitor electrodes for enhanced charge storage, RSC Adv. 5 (2015) 58100−58106. https://doi.org/10.1039/C5RA11597C

[66] P. Wen, P. Gong, J. Sun, J. Wang, S. Yang, Design and synthesis of Ni-MOF/CNT composites and rGO/carbon nitride composites for an asymmetric supercapacitor with high energy and power density, J. Mater. Chem. A 3 (2015), 13874−13883. https://doi.org/10.1039/C5TA02461G

[67] X. Xu, J. Tang, H. Qian, S. Hou, Y. Bando, M.S.A. Hossain, L. Pan, Y. Yamauchi, Three-dimensional networked metal-organic frameworks with conductive polypyrrole tubes for flexible supercapacitors, ACS Appl. Mater. Interfaces 9 (2017) 38737−38744. https://doi.org/10.1021/acsami.7b09944

[68] P. Srimuk, S. Luanwuthi, A. Krittayavathananon, M. Sawangphruk, Solid-type supercapacitor of reduced grapheme oxide-metal organic framework composite coated on carbon fiber paper. Electrochim. Acta 157 (2015) 69−77. https://doi.org/10.1016/j.electacta.2015.01.082

[69] K. Jayaramulu, M. Horn, A. Schneemann, H. Saini, A. Bakandritsos, V. Ranc, M. Petr, V. Stavila, C. Narayana, B. Scheibe, S. Kment, M. Otyepka, N. Motta, D. Dubal, R. Zboril, R.A.Fischer, Covalent graphene-MOF hybrids for high-performance asymmetric supercapacitors, Adv. Mater. 33 (2021), 2004560. https://doi.org/10.1002/adma.202004560

[70] X. Cao, L. Cui, B. Liu, Y. Liu, D. Jia, W. Yang, J.M. Razal, J. Liu, Reverse synthesis of star anise-like cobalt doped Cu-MOF/Cu2+1O hybrid materials based on a Cu(OH)2 precursor for high performance supercapacitors, J. Mater. Chem. A 7 (2019) 3815−3827. https://doi.org/10.1039/C8TA11396C

[71] P.Banerjee, D.E. Lobo, R. Middag, W. Ng, M.E. Shaibani, M. Majumder, Electrochemical capacitance of Ni-doped metal organic framework and reduced graphene oxide composites: More than the sum of its parts, ACS Appl. Mater. Interfaces 7 (2015) 3655−3664. https://doi.org/10.1021/am508119c

[72] D.Tian, N. Song, M. Zhong, X. Lu, C. Wang, Bimetallic MOF nanosheets decorated on electrospun nanofibers for high performance asymmetric supercapacitors, ACS Appl. Mater. Interfaces 12 (2020) 1280−1291. https://doi.org/10.1021/acsami.9b16420

[73] Y. Jiao, G. Chen, D. Chen, J. Pei, Y. Hu, Bimetal-organic framework assisted polymerization of pyrrole involving air oxidant to prepare composite electrodes for portable energy storage, J. Mater.Chem. A 5 (2017) 23744−23752. https://doi.org/10.1039/C7TA07464F

[74] C.Wang, C. Liu, J. Li, X. Sun, J. Shen, W. Han, L.Wang, Electrospun metal- organic framework derived hierarchical carbon nanofibers with high performance for supercapacitors, Chem. Commun. 53 (2017) 1751-1754. https://doi.org/10.1039/C6CC09832K

[75] X.Cao, C. Tan, M. Sindoro, H. Zhang, Hybrid micro-/nano- structures derived from metal-organic frameworks: preparation and applications in energy storage and conversion, Chem. Soc. Rev. 46 (2017) 2660-2677. https://doi.org/10.1039/C6CS00426A

[76] F. Bonaccorso, L. Colombo, G. Yu, M. Stoller, V. Tozzini, A.C. Ferrari, R.S.Ruoff, V.Pellegrini, Graphene, related two-dimensional crystals, and hybrid systems for energy conversion and storage, Science. 347 (2015) 1246501. https://doi.org/10.1126/science.1246501

[77] S. Dai, Y. Yuan, J. Yu, J. Tang, J. Zhou, W. Tang, Metal-organic framework-templated synthesis of sulfur-doped core-sheath nanoarrays and nanoporous carbon for flexible all-solid-state asymmetric supercapacitors, Nanoscale 10 (2018) 15454-15461. https://doi.org/10.1039/C8NR03743D

[78] M. Gao, J. Fu, M. Wang, K. Wang, S. Wang, Z. Wang, Z.Chen, Q.Xu, A self- template and self-activation co-coupling green strategy to synthesize high surface area ternary-doped hollow carbon microspheres for high performance supercapacitors, J. Colloid Interf. Sci. 524, (2018). 165-176. https://doi.org/10.1016/j.jcis.2018.04.027

[79] W. Ma, L. Xie, L.Dai, G. Sun, J. Chen, F. Su, Y.Cao, H.Lei, O.Kong, C.M.Chen, Influence of phosphorus doping on surface chemistry and capacitive behaviors of porous carbon electrode, Electrochim. Acta. 266 (2018) 420-430. https://doi.org/10.1016/j.electacta.2018.02.031

[80] D.K.Kim, S. Bong, X. Jin, K.D. Seong, M, Hwang, N.D. Kim, N. H.You, Y.Piao,Facile in situ synthesis of multiple-heteroatom-doped carbons derived from polyimide precursors for flexible all-solid-state supercapacitors, ACS Appl. Mater. Inter. 11, (2019).1996-2005. https://doi.org/10.1021/acsami.8b15162

[81] B. Liu, H. Shioyama, T. Akita, Q. Xu, Metal-Organic Framework as a template for porous carbon synthesis, J. Am. Chem. Soc. 130 (2008) 5390-5391. https://doi.org/10.1021/ja7106146

[82] B. Liu, H. Shioyama, H. Jiang, X. Zhang, Q. Xu, Metal-organic framework (MOF) as a template for syntheses of nanoporous carbons as electrode materials for supercapacitor Carbon 48 (2010) 456-463. https://doi.org/10.1016/j.carbon.2009.09.061

[83] J. Hu, H. Wang, Q. Gao, H. Guo, Porous carbons prepared by using metal-organic framework as the precursor for supercapacitors Carbon 48 (2010) 3599-3606. https://doi.org/10.1016/j.carbon.2010.06.008

[84] H.B. Aiyappa, P. Pachfule, R. Banerjee, S. Kurungot, Porous carbons from nonporous MOFs: influence of ligand characteristics on intrinsic properties of end carbon, Cryst. Growth Des. 13 (2013) 4195-4199. https://doi.org/10.1021/cg401122u

[85] H.L. Jiang, B. Liu, Y.Q. Lan, K. Kuratani, T. Akita, H. Shioyama, F. Zong, Q. Xu, From metal-organic framework to nanoporous carbon: toward a very high surface area and hydrogen uptake J.Am. Chem. Soc. 133 (2011) 11854-11857. https://doi.org/10.1021/ja203184k

[86] W. Chaikittisilp, M. Hu, H. Wang, H.S. Huang, T. Fujita, K.C. Wu, L.C. Chen, Y. Yamauchi, K. Ariga, Nanoporous carbons through direct carbonization of a zeolitic imidazolate framework for supercapacitor electrodes, Chem. Commun. 48 (2012) 7259-7261. https://doi.org/10.1039/c2cc33433j

[87] A.J. Amali, J.K. Sun, Q. Xu, From assembled metal-organic framework nanoparticles to hierarchically porous carbon for electrochemical energy storage, Chem. Commun. 50 (2014) 1519-1522. https://doi.org/10.1039/C3CC48112C

[88] R.R. Salunkhe, Y. Kamachi, N.L. Torad, S.M. Hwang, Z. Sun, S.X. Dou, J.H. Kim, Y. Yamauchi, Fabrication of symmetric supercapacitors based on MOF-derived nanoporous carbons, J. Mater. Chem. A 2 (2014) 19848-19854. https://doi.org/10.1039/C4TA04277H

1 [89] J. Tang, R.R. Salunkhe, J. Liu, N.L. Torad, M. Imura, S. Furukawa, Y. Yamauchi, Thermal conversion of core-shell metal-organic frameworks: a new method for selectively functionalized nanoporous hybrid carbon, J.Am. Chem. Soc. 137 (2015) 1572-1580. https://doi.org/10.1021/ja511539a

[90] A. Banerjee, K.K. Upadhyay, D. Puthusseri, V. Aravindan, S. Madhavi, S. Ogale, MOF-derived crumpled-sheet-assembled perforated carbon cuboids as highly effective cathode active materials for ultra-high energy density Li-ion hybrid electrochemical capacitors (Li-HECs) Nanoscale 6 (2014) 4387-4394. https://doi.org/10.1039/c4nr00025k

[91] T.H.Chang, C. Young, M.H. Lee, R.R. Salunkhe, S.M. Alshehri, T. Ahamad, T.Islam, C.W.Wu, S.A. Hossain, Y.Yamauchi, K.C.Ho, Synthesis of MOF-525 derived nanoporous carbons with different particle sizes for supercapacitor application, Chem. Asian J. 12, (2017) 2857-2862. https://doi.org/10.1002/asia.201701082

[92] Y. Wang, X. Xie, B. Zhang, J. Luo, S. Wang, S. Nie, S. Lin, H.Yang, A New Cd- based metal organic framework derived nitrogen doped nano-porous carbon for high supercapacitor performance, Polyhedron. 189, (2020) 114726. https://doi.org/10.1016/j.poly.2020.114726

[93] S.C. Wu, P.H. Chang, S.H. Chou, C.Y. Huang, T.C. Liu, C.H Peng, Waffle-like carbons combined with enriched mesopores and highly heteroatom-doped derived from sandwiched MOF/LDH/MOF for high-rate supercapacitor. Nanomaterials 10. (2020) 2388 https://doi.org/10.3390/nano10122388

[94] M. Liu, F. Zhao, D. Zhu, H. Duan, Y. Lv, L. Li, L.Gan, Ultramicroporous carbon nanoparticles derived from metal-organic framework nanoparticles for high-performance supercapacitors, Mater. Chem. Phys. 211 (2018) 234-241. https://doi.org/10.1016/j.matchemphys.2018.02.030

[95] W. Bao, A.K. Mondal, J. Xu, C. Wang, D. Su, G. Wang, G. 3D Hybrid- porous carbon derived from carbonization of metal organic frameworks for high performance supercapacitors, J. Power Sourc. 325 (2016) 286-291. https://doi.org/10.1016/j.jpowsour.2016.06.037

[96] Z. Tang, G. Zhang, H. Zhang, L. Wang, H. Shi, D. Wei, H.Duan, MOF derived N-doped carbon bubbles on carbon tube arrays for flexible high- rate supercapacitors, Energ. Storage Mater. 10, (2018) 75-84. https://doi.org/10.1016/j.ensm.2017.08.009

[97] M.L. Yue, C.Y. Yu, H.H. Duan, B.L. Yang, X.X. Meng, Z.X. Li, six isomorphous window-beam MOFs: Explore the effects of metal ions on mof-derived carbon for supercapacitors, Chem. Eur. J. 24 (2018) 16160-16169. https://doi.org/10.1002/chem.201803554

[98] J. Cai, Y. Song, X. Chen, Z. Sun, Y. Yi, J. Sun, Q. Zhang, MOF-derived conductive carbon nitrides for separator-modified li-s batteries and flexible supercapacitors, J. Mater. Chem. A. 8 (2020) 1757-1766. https://doi.org/10.1039/C9TA11958B

[99] Y. Chen, L. Hu, Novel Co3O4 porous polyhedrons derived from metal-organic framework toward high performance for electrochemical energy devices, J. Solid State Chem. 239 (2016) 23-29. https://doi.org/10.1016/j.jssc.2016.02.009

[100] K. Wang, X. Yi, X. Luo, Y. Shi, J. Xu, Fabrication of Co3O4 pseudocapacitor electrodes from nanoscale cobalt-organic frameworks, Polyhedron 109 (2016) 26-32. https://doi.org/10.1016/j.poly.2016.01.046

[101] W. Xu, T.T. Li, Y.Q. Zheng, Porous Co3O4 nanoparticles derived from a Co(II)-cyclohexanehexacarboxylate metal-organic framework and used in a supercapacitor with good cycling stability, RSC Adv. 6 (2016) 86447-86454 https://doi.org/10.1039/C6RA17471J

[102] Z. Sun, F. Huang, Y. Sui, F. Wei, J. Qi, Q. Meng, H. Hu, Y. He, Cobalt oxide composites derived from zeolitic imidazolate framework for high-performance supercapacitor electrode, J. Mater. Sci. Mater. Electron. 28 (2017) 14019-14025 https://doi.org/10.1007/s10854-017-7252-4

[103] H. Li, F. Yue, C. Yang, P. Qiu, P. Xue, Q. Xu, J. Wang, Porous nanotubes derived from a metal-organic framework as high-performance supercapacitor electrodes, Ceram. Int. 42 (2016) 3121-3129. https://doi.org/10.1016/j.ceramint.2015.10.101

[104] C. Guan, W. Zhao, Y. Hu, Z. Lai, X. Li, S. Sun, H. Zhang, A.K. Cheetham, J. Wang, Cobalt oxide and N-doped carbon nanosheets derived from a single two-dimensional metal-organic framework precursor and their application in flexible asymmetric supercapacitors, Nanoscale Horiz. 2 (2017) 99-105. https://doi.org/10.1039/C6NH00224B

[105] Z. Wang, Y. Liu, C. Gao, H. Jiang, J. Zhang, A porous Co(OH)2 material derived from a MOF template and its superior energy storage performance for supercapacitors, J. Mater. Chem. A 3 (2015) 20658-20663. https://doi.org/10.1039/C5TA04663G

[106] T. Deng, Y. Lu, W. Zhang, M. Sui, X. Shi, D. Wang, W. Zheng, Inverted design for high-performance supercapacitor via Co(OH)2-derived highly oriented mof electrodes, Adv. Energy Mater. 8, (2017) 1702294. https://doi.org/10.1002/aenm.201702294

[107] M.K. Wu, C. Chen, J.J. Zhou, F.Y. Yi, K. Tao, L. Han, MOF-derived hollow double-shelled NiO nanospheres for high-performance supercapacitors, J. Alloys Compd., 734 (2018)1-8. https://doi.org/10.1016/j.jallcom.2017.10.171

[108] S. Chen, D. Cai, X. Yang, Q. Chen, H. Zhan, B. Qu, T. Wang, Metal-organic frameworks derived nanocomposites of mixed-valent mnox nanoparticles in-situ grown on ultrathin carbon sheets for high-performance supercapacitors and lithium-ion batteries, Electrochim. Acta 256 (2017) 63-72. https://doi.org/10.1016/j.electacta.2017.10.016

[109] Y. Zhang, B. Lin, J. Wang, P. Han, T. Xu, Y. Sun, X. Zhang, H. Yang, Polyoxometalates@metal-organic frameworks derived porous moo3@cuo as electrodes for symmetric all-solid-state supercapacitor, Electrochim. Acta 191(2016) 795-804. https://doi.org/10.1016/j.electacta.2016.01.161

[110] J. Tian, B. Lin, Y. Sun, X. Zhang, H. Yang, Porous WO3@CuO composites derived from polyoxometalates@metal organic frameworks for supercapacitor, Mater. Lett. 206 (2017) 91-94. https://doi.org/10.1016/j.matlet.2017.06.116

[111] S. Maiti, A. Pramanik, S. Mahanty, Extraordinarily high pseudocapacitance of metal organic framework derived nanostructured cerium oxide, ChemComm. 50 (2014) 11717-11720. https://doi.org/10.1039/C4CC05363J

[112] W. Meng, W. Chen, L. Zhao, Y. Huang, M. Zhu, Y. Huang, Y. Fu, F. Geng, J. Yu, X. Chen, C. Zhi, Porous Fe3O4/carbon composite electrode material prepared from metal-organic framework template and effect of temperature on its capacitance, Nano Energy 8 (2014) 133-140. https://doi.org/10.1016/j.nanoen.2014.06.007

[113] S. Ullah, I.A. Khan, M. Choucair, A. Badshah, I. Khan, M.A. Nadeem, A novel Cr2O3-carbon composite as a high performance pseudo-capacitor electrode material,

Electrochim. Acta 171 (2015) 142-149. https://doi.org/10.1007/s10440-015-0021-6

[114] I.A. Khan, A. Badshah, M.A. Nadeem, N. Haider, M.A. Nadeem, A copper based metal-organic framework as single source for the synthesis of electrode materials for high-performance supercapacitors and glucose sensing applications, Int. J. Hydrogen Energy 39 (2014) 19609-19620. https://doi.org/10.1016/j.ijhydene.2014.09.106

[115] Y. Zhang, H. Chen, C. Guan, Y. Wu, C. Yang, Z. Shen, Q. Zou, Energy-saving synthesis of mof-derived hierarchical and hollow co(vo3)2-co(oh)2 composite leaf arrays for supercapacitor electrode materials, ACS. Appl. Mater. Interfaces 10 (2018) 18440-18444. https://doi.org/10.1021/acsami.8b05501

[116] C. Guan, X. Liu, W. Ren, X. Li, C. Cheng, J. Wang, rational design of metal-organic framework derived hollow NiCo2O4 arrays for flexible supercapacitor and electrocatalysis, Adv. Energy Mater. 7 (2017)1602391. https://doi.org/10.1002/aenm.201770086

[117] Q. Yang, Q. Wang, Y. Long, F. Wang, L. Wu, J. Pan, J. Han, Y. Lei, W. Shi, S. Song, In situ formation of Co9S8 quantum dots in mof-derived ternary metal layered double hydroxide nanoarrays for high-performance hybrid supercapacitors, Adv. Energy Mater. 10 (2020) 1903193. https://doi.org/10.1002/aenm.201903193

[118] Y. Huang, L. Quan, T. Liu, Q. Chen, D. Cai, H. Zhan, Construction of MOF-derived hollow Ni-Zn-Co-S nanosword arrays as binder-free electrodes for asymmetric supercapacitors with high energy density, Nanoscale 10 (2018)14171-14181. https://doi.org/10.1039/C8NR03919D

[119] A. Jayakumar, R.P. Antony, R. Wang, J.M. Lee, MOF-Derived Hollow Cage NixCo3−xO4 and Their Synergy with Graphene for Outstanding Supercapacitors, Small 13 (2017) 1603102. https://doi.org/10.1002/smll.201603102

[120] T. Xu, G. Li, L. Zhao, Ni-Co-S/Co(OH)2 nanocomposite for high energy density all-solid-state asymmetric supercapacitors, Chem. Eng. J. 336 (2018) 602-611. https://doi.org/10.1016/j.cej.2017.12.065

[121]W. Li,B. Zhang, R. Lin, S. Ho-Kimura, G. He, X. Zhou, J. Hu, I.P. Parkin, A dendritic nickel cobalt sulfide nanostructure for alkaline battery electrodes, Adv. Funct. Mater. 28 (2018) 1705937. https://doi.org/10.1002/adfm.201705937

[122] Tao, K.; Han, X.; Cheng, Q.; Yang, Y.; Yang, Z.; Ma, Q.; Han, L. A zinc cobalt sulfide nanosheet array derived from a 2D bimetallic metal-organic frameworks for high-performance supercapacitors, Chem.-A Eur. J. 24, (2018) 12584-12591. https://doi.org/10.1002/chem.201800960

Emerging Materials for Next Frontier Energy and Environment Applications　　Materials Research Forum LLC
Materials Research Foundations 170 (2024) 94-116　　　　　https://doi.org/10.21741/9781644903292-5

[123] H. Hu, B. Y. Guan, W. Xiong Lou, Construction of complex CoS hollow structures with enhanced electrochemical properties for hybrid supercapacitors, Chem 1 (2016) 102-113. https://doi.org/10.1016/j.chempr.2016.06.001

[124] F. Cao, M. Zhao, Y. Yu, B. Chen, Y. Huang, J. Yang, X. Cao, Q. Lu, X. Zhang, Z. Zhang, C. Tan, H.Zhang, Synthesis of Two-dimensional CoS1.097/nitrogen-doped carbon nanocomposites using metal-organic framework nanosheets as precursors for supercapacitor application, J. Am. Chem. Soc., 138 (2016) 6924-6927. https://doi.org/10.1021/jacs.6b02540

[125] S. Liu, M. Tong, G. Liu, X. Zhang, Z. Wang, G. Wang, W. Cai, H. Zhang, H. Zhao, S,N-containing Co-MOF derived Co9S8@S,N-doped carbon materials as efficient oxygen electrocatalysts and supercapacitor electrode materials, Inorg. Chem. Front. 4 (2017) 491-498 https://doi.org/10.1039/C6QI00403B

[126] S. Zhang, D. Li, S. Chen, X. Yang, X. Zhao, Q. Zhao, S. Komarneni, D. Yang, Highly stable supercapacitors with MOF-derived Co9S8/carbon electrodes for high rate electrochemical energy storage, J. Mater. Chem. A 5 (2017) 12453-12461. https://doi.org/10.1039/C7TA03070C

[127] J.S. Chen, C. Guan, Y. Gui, D.J. Blackwood, Rational Design of Self-Supported Ni3S2 Nanosheets Array for Advanced Asymmetric Supercapacitor with a Superior Energy Density, ACS Appl. Mater. Interfaces 9 (2017) 496-504. https://doi.org/10.1021/acsami.6b14746

[128] C. Qu, L. Zhang, W. Meng, Z. Liang, B. Zhu, D. Dang, S. Dai, B. Zhao, H. Tabassum, S. Gao, H. Zhang, W. Guo, R. Zhao, X. Huang, M. Liu, R. Zou, MOF-derived α-NiS nanorods on graphene as an electrode for high-energy-density supercapacitors, Journal of Materials Chemistry A, 6 (2018) 4003-4012. https://doi.org/10.1039/C7TA11100B

[129] G.-C. Li, M. Liu, M.-K. Wu, P.-F. Liu, Z. Zhou, S.-R. Zhu, R. Liu, L. Han, MOF-derived self-sacrificing route to hollow NiS2/ZnS nanospheres for high performance supercapacitors, RSC Adv., 6 (2016)103517-103522. https://doi.org/10.1039/C6RA23071G

[130] X.Y. Yu, L. Yu, B. Wu Hao, W. Lou Xiong, Formation of nickel sulfide nanoframes from metal-organic frameworks with enhanced pseudocapacitive and electrocatalytic properties, Angew. Chem. 127 (2015) 5421-5425. https://doi.org/10.1002/ange.201500267

[131] P. Cai, T. Liu, L. Zhang, B. Cheng, J. Yu, ZIF-67 derived nickel cobalt sulfide hollow cages for high-performance supercapacitors, Appl. Surf. Sci. 504 (2020) 144501. https://doi.org/10.1016/j.apsusc.2019.144501

[132] D. Tian, S. Chen, W. Zhu, C. Wang, X. Lu, Metal-organic framework derived hierarchical Ni/Ni3S2 decorated carbon nanofibers for high-performance supercapacitors, Mater. Chem. Front. 3, (2019) 1653-1660. https://doi.org/10.1039/C9QM00296K

[133] L. Li, J. Zhao, Y. Zhu, X. Pan, H. Wang, J. Xu, Bimetallic Ni/Co-ZIF-67 derived NiCo2Se4/N-doped porous carbon nanocubes with excellent sodium storage performance, Electrochim. Acta 353, (2020) 136532. https://doi.org/10.1016/j.electacta.2020.136532

[134] L. Tan, D. Guo, D. Chu, J. Yu, L. Zhang, J. Yu, J. Wang, Metal organic frameworks template-directed fabrication of hollow nickel cobalt selenides with pentagonal structure for high-performance supercapacitors, J. Electroanal. Chem. 851, (2019) 113469. https://doi.org/10.1016/j.jelechem.2019.113469

[135] L.Quan, T.Liu, M. Yi, Q. Chen, D. Cai, H. Zhan, Construction of hierarchical nickel cobalt selenide complex hollow spheres for pseudocapacitors with enhanced performance, Electrochim. Acta 281, (2018) 109-116. https://doi.org/10.1016/j.electacta.2018.05.100

[136]C. Miao, C. Zhou, H.E. Wang, K. Zhu, K. Ye, Q. Wang, J. Yan, D. Cao, N. Li, G. Wang, Hollow Co-Mo-Se nanosheet arrays derived from metal-organic framework for high-performance supercapacitors, J. Power Sources 490 (2021) 229532. https://doi.org/10.1016/j.jpowsour.2021.229532

[137] B.Ameri, A.M. Zardkhoshoui, S.S.H. Davarani, Metal-organic-framework derived hollow manganese nickel selenide spheres confined with nanosheets on nickel foam for hybrid supercapacitors, Dalton Trans. 50 (2021) 8372-8384. https://doi.org/10.1039/D1DT01215K

[138] Q. He, T. Yang, X. Wang, P. Zhou, S. Chen, F. Xiao, P. He, L. Jia, T. Zhang, D. Yang, Metal-organic framework derived hierarchical zinc nickel selenide/nickel hydroxide microflower supported on nickel foam with enhanced electrochemical properties for supercapacitor, J. Mater. Sci. Mater. Electron. 32 (2021) 3649-3660. https://doi.org/10.1007/s10854-020-05111-x

[139] C.Miao, X.Xiao, Y.Gong, K. Zhu, K. Cheng, K. Ye, J. Yan, D. Cao, G. Wang, P. Xu, Facile synthesis of metal-organic framework-derived CoSe2 nanoparticles embedded in the N-doped carbon nanosheet array and application for supercapacitors, ACS Appl. Mater. Interfaces 12 (2020) 9365-9375. https://doi.org/10.1021/acsami.9b22606

[140] W. Chu, Y. Hou, J. Liu, X. Bai, Y. f. Gao, Z. Cao, Zn-Co phosphide porous nanosheets derived from metal-organic-frameworks as battery-type positive electrodes for high-performance alkaline supercapacitors, Electrochim. Acta 364, (2020) 137063. https://doi.org/10.1016/j.electacta.2020.137063

[141] Z. Lv, Q. Zhong, Y. Bu, In-situ conversion of rGO/Ni2P composite from GO/Ni-MOF precursor with enhanced electrochemical property, Appl. Surf. Sci. 439 (2018) 413-419. https://doi.org/10.1016/j.apsusc.2017.12.185

[142] X. Liu, W. Ang, C. Guan, L. Zhang, Y. Qian, A. M. Elshahawy, D. Zhao, S. J. Pennycook, J. Wang, Ni-doped cobalt-cobalt nitride heterostructure arrays for high-power supercapacitors ACS Energy Lett. 3 (2018) 2462-2469. https://doi.org/10.1021/acsenergylett.8b01393

Emerging Materials for Next Frontier Energy and Environment Applications   Materials Research Forum LLC
Materials Research Foundations 170 (2024) 117-144   https://doi.org/10.21741/9781644903292-6

Chapter 6

Transition metal oxides-MXene nanocomposite: The next frontier in supercapacitors

Jayachandran Madhavan, Pavithra Karthikesan, Harshini Sharan, Alagiri Mani*

Department of Physics and Nanotechnology, SRM Institute of Science and Technology, Kattankulathur, Chengalpattu, Tamil Nadu - 603 203, India

* alagirim@srmist.edu.in

Abstract

The steady depletion of non-renewable energy sources along with global warming, has emphasized environment friendly energy systems worldwide. Consequently, there has been a significant increase in demand for efficient energy storage devices, particularly for highly efficient energy storage devices such as supercapacitors and secondary batteries. Comparing with energy storage devices supercapacitor possess a very high-power density, a decent energy density and have an excellent cyclic stability. However, their limited energy density restricts them from being integrated with our daily energy needs. Electrode material based on transition metal oxide (TMO) are particularly intriguing due to their exceptional blend of structural, mechanical, electrical, and electrochemical capabilities. TMO are promising electrode materials for supercapacitors owing to its high capacitance and energy density attributed to rich redox chemistry, as well as their high reversibility, rapid charge-discharge operations, minimal expense owing to availability, and ecological sustainability. However, the significant obstacles that need to be surmounted are inadequate electrical conductivity, rate capability, poor cycle life, and low power density. To overcome these hindrances nanocomposite of TMO with 2D layered materials such as MXenes which provides high electronic conductivity and large surface area for better activation of TMO to enhance the charge storage capabilities. This chapter systematically aims on the most recent developments in MXene-TMO heterostructure electrode materials for supercapacitors and highlights their merits.

Keywords

MXene, Transition Metal Oxides, Two-Dimensional Materials, Supercapacitors, Energy Storage

Contents

Transition metal oxides-MXene nanocomposite: The next frontier in supercapacitors 117

1.   Introduction 118

2.   Energy storage systems 119

3.   Types of energy storage systems 120

4.   Electrical and electrochemical energy storage systems 121

5.   Supercapacitors 123

        5.1   Electric double layer capacitors (EDLCs) 124

        5.2   Pseudocapacitors 124

        5.3   Hybrid supercapacitors 125

        5.3.1   Asymmetric supercapacitors 125

        5.3.2   Composite supercapacitors 125

        5.3.3   Battery-type Supercapacitors 125

        5.4   Electrolytes for supercapacitors 126

        5.4.1   Liquid electrolytes 126

        5.4.2   Polymer-based electrolytes 127

        5.4.3   Redox electrolytes 127

6.   MXenes 127

        6.1   Synthesis of MXenes 128

        6.1.1   HF etching 129

        6.1.2   Urea glass route 129

        6.1.3   Molten salt etching 129

        6.1.4   Chemical vapor deposition 130

        6.1.5   Hydrothermal synthesis 130

        6.1.6   Electrochemical synthesis 130

        6.3   MXene properties 131

7.   Transition metal oxides for supercapacitors 132

8.   MXene-transition metal oxide nanocomposite for supercapacitors 133

Conclusion 137

References 138

1.   Introduction

After the Industrial Revolution, the world nations socio-economic development has depended extensively on fossil fuels (such as petroleum, coal and oil) as a readily available power supply to fulfil modern humankind's energy demand [1,2]. As of 2012, total global energy consumption via fossil fuels was predicted to be 13.731 billion tonnes of oil equivalent (BTOE), with a projected increase to 18.30 BTOE by 2035 [3]. The continuous usage of fossil fuels affects the environment and it's living beings by means of global warming, air pollution, water pollution and so on. If this trend continues, global temperature of 1.5°C will be increased at the end of 2052 [4]. With the predicted world population increase by 26% at the end of 2050, to cope up with the increasing global demand for energy and also to ensure the sustainability of our planet, a minimum of 10 terawatts (TWs) of green power generating capacity has to be established by the end of 2050 [5,6]. Renewable energy sources such as solar, wind, tidal, and so on are the sought-after non-conventional, sustainable, and eco-friendly energy sources to cater to modern day energy requirement. The drawback of using these energy sources is that these are intermittent energy sources. The Solar power cannot be harvested during cloudy and nighttime, wind energy cannot be extracted in the absence of steady wind and so on. To address this issue, energy storage system comes into play. Coupling energy storage systems with the renewable energy sources will provide an uninterrupted power supply.

An ideal energy storage system should store and deliver energy instantaneously when required. In this regard, the widely used energy storage systems are the secondary batteries. The drawback of using these secondary batteries is due to their low power density, poor cycle stability and risk of explosion. To overcome these barriers, supercapacitors come into play which has high power density for fast charging and discharging, better cycle stability and has no risk of explosion. In the upcoming sections, different types of energy storage, supercapacitor working and its types, state-of-art electrode materials for supercapacitors, and future prospects is elaborated in detail.

2.   Energy storage systems

The Energy storage systems (ESS) play an important part in integration of renewable energy sources into the grid. Modern power demand with fluctuating renewable energy sources can be only catered with the help of mediating storage system. To utilize the maximum extent of renewable energy, The renewable sources should be effectively tapped when it is available; otherwise, their potential is lost. ESS assist in translating this potential to renewable systems in response to fluctuation in grid frequency, allowing the power system to quickly recover from unexpected imbalances between electricity demand and supply [7,8].

Solar and wind are the widely sought after renewable sources of energy, although they both have challenges with intermittent and unpredictable availability, as previously discussed. Interconnecting renewable systems over a wide expanse allows for efficient management of the fluctuation in their energy sources. Furthermore, modern power electronics helps in achieving a perfect synchronization with the grid for providing a constant power supply. ESS facilitates adaptable functioning and seamless integration of power generators with electric grid in such scenarios. These systems enhance the efficiency of distributing wind or solar electricity produced on-site and can be used during periods of high demand [9]. The development of contemporary power grids, such as smart grids, microgrids and nanogrids, relies heavily on the assistance of ESS. These systems play a critical role in providing backup and auxiliary reserve for the power sources involved. The necessity for ESS appears in several aspects of power management, including creating power backups, auxiliary supports, demand control and seasonal storage.

There exists a broad range of ESS, each with unique characteristics that make them well-suited for certain applications in contemporary power systems. Conversely, the ESS importance in future power grids has not been recognized and valued yet.

3.   Types of energy storage systems

The ESS are mainly classified by their energy storage mechanism like, thermal, mechanical, electrical, and electrochemical. Fig. 1. Shows different energy storage systems and brief outline of these energy storage systems are as discussed:

Flywheel Energy storage system: As the name suggests, mechanical flywheels are utilized in this energy storage system. Flywheels are rotating devices which stores energy in the form of kinetic energy. The energy is contained inside a rapidly spinning rotor, which is designed as a large rotating cylinder. During the charging phase, the rotor undergoes rapid acceleration, reaching speeds ranging up to 50,000 rpm. The flywheel stores energy by maintaining a consistent rotational velocity of the spinning body. During the discharge process, the rotor slows down and operates the machine which it is connected to. The benefits of using flywheel system are due to its low maintenance, low cost, high efficiency, and long life. Although with these advantages, applications requiring extended periods of energy storage is not suitable due to its inability to retain the stored energy effectively, mostly due to its breaking losses [10].

Compressed Air Energy Storage System: This energy storage system utilizes natural caverns beneath the surface or man-made constructed air chambers. There are two types of Compressed air energy storage: Adiabatic and Diabatic. The primary cause of reduced efficiency is due to the heat generation during the compression of air and thus it is diabatic in nature. When the thermal energy is conserved and then used for decompression, then the system is considered adiabatic, and this exhibits significantly improved efficiency. The air is compressed using an electric compressor and stored in the underground air chambers. The advantages of using these systems are they have large energy and power rating, longer lifetime and small energy loss. However, the major drawback is that it is region specific, cannot be used in a small scale (only suitable for grid level application), requires extensive funding [11].

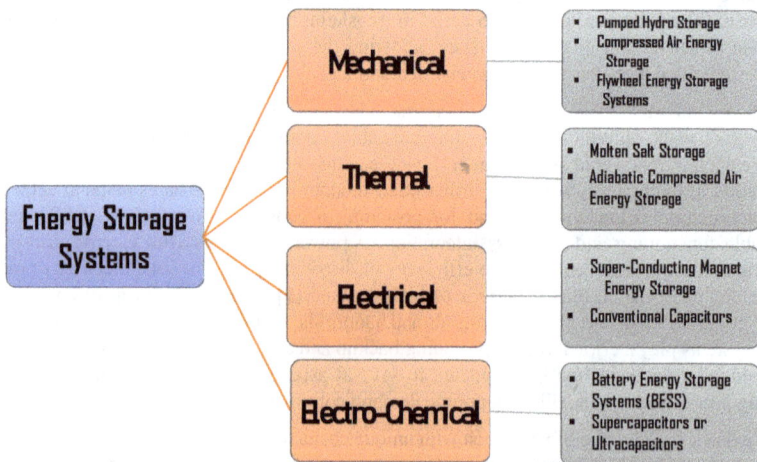

Figure 1. Different energy storage systems

Pumped Hydro Storage: This is an established and extensively used method of energy storage for grid and power applications, several water reservoirs are used for Pumped Hydro Storage projects, which seek to save energy during periods of energy surplus. This is achieved by using a pump to raise water to higher elevations and then producing electricity by reversing the process with the assistance of turbine. The benefits of these systems are their substantial energy and power capacity, extended lifespan, impressive efficiency and minimal discharge losses. The primary obstacle with this kind of storage is similar to that of Compressed Air Energy Storage system which is region specific and long development period. Several concepts are currently being explored and developed in this energy storage system [12].

Superconducting Magnet Energy Storage: The system is comprised of a superconductive coil, refrigeration system, power system and a vacuum environment. The energy is stored in the magnetic field generated via a current flowing through the superconducting coil. The benefits of this system include quick response, immediate power accessibility, and can be scaled down and utilized according to the energy demands. These storage systems are now in the development phase and are expected to become very desirable energy storage solutions for power grid applications.

4.   Electrical and electrochemical energy storage systems

The extensively used electrical and electrochemical energy storage systems are capacitors, secondary batteries, and supercapacitors. These device's power and energy densities are plotted in ragone plot which is depicted in Fig. 2. and these energy storage systems are briefed below:

Secondary Batteries:

The core unit of rechargeable or secondary batteries is the electrochemical cell, which has an anode, a cathode and a electrolyte impregnated separator. The separator acts as a barrier to prevent short-circuit between anode and cathode. During electrochemical oxidation, i.e. discharge at anode (negative electrode) the electrons are produced and flow into external circuit.

$$A \rightarrow A^{z+} + ze^- \qquad \text{During oxidation} \qquad (1)$$

Then electrons enter the electrolyte through the electrochemical reduction, at the cathode (positive electrode) as positive and negative ions.

$$C + ze^- \rightarrow C^{z-} \qquad \text{During reduction} \qquad (2)$$

The anions C^{z-} move in the electrolyte and carry the current from the cathode (positive electrode) back to anode (negative electrode) to complete the electrical circuit. Similarly, the cations A^{z+} move from anode to cathode (positive electrode). During the charging process the electrode reactions are reversed (electron flow in opposite direction) and the battery is restored to its initial state [13].

There are several kinds of batteries with distinct features that may be used for a wide range of applications, from electrical vehicles to support backup systems for power grids.

Lead acid batteries: This battery consists of PbO_2 as positive electrode and Pb sponge as negative electrode. Both electrodes are submerged in a sulphuric acid electrolyte. These batteries suffer from maintenance problems related to the electrolyte and corrosion of the electrodes. Lately, liquid electrolytes are substituted with gel-based electrolytes. Also, the efficiency can be enhanced by utilizing nanostructured materials. The advantage of lead acid batteries includes effortless installation, cost-effective, high retention, and high efficiency. The drawbacks are the maintenance concerns, low power density, early failure when run with partial discharge and not environmentally benign [14].

Lithium-ion batteries: It is a rechargeable electrochemical battery that facilitates the transportation of lithium ions between the negative and positive electrodes via an electrolyte and vice versa. The cathode and anode are physically separated with a separator, often consisting of a thin polyolefin film with electrolyte. The cathode and anode hosts lithium ions in them; During discharge process, the lithium ions move from the anode and gets intercalated within the crystallographic structure of the cathode. While charging, this process proceeds in reverse direction. In commercial applications, a single lithium-ion cell has an operating voltage range of around 3 V to 4 V. The drawback of Lithium-ion batteries is that they use organic solvents as electrolytes which is volatile and flammable, during cell overheating the integrity of the film may be compromised leading to a cell failure [15].

Flow batteries: Flow battery technologies have been drawing significant attention for large scale power applications since the early 1990s. Flow batteries use chemical oxidation-reduction (redox) reactions to produce electrical energy. Two electrolyte solutions, namely the catholyte and the anolyte, are contained in separate storage tanks that are parted by a membrane. The redox means the contrasting valence charge states on opposite sides of the membrane that uses as electrical potential. As the electrolyte injected to the membrane, exchange of ions causes the generation of an electric current, where charge is either withdrawn or supplied via two electrodes. Iron and vanadium-based electrolytes are predominantly used for redox flow batteries. If the anolyte and catholyte consist of distinct variations of the same metal, there is negligible chance for cross contamination. Unlike vanadium and iron flow batteries, with similar electrolytes on either side of the membrane, the hybrid redox flow batteries use two distinct electrolytes. Considering zinc-bromide, a significant amount of energy is stored onto the anode plates by depositing zinc metal during the charging process. In the case of vanadium flow battery, the amount of energy it can store depends on the size of the tank, while the output power is determined by the active surface area of the anode. These two factors are not connected to each other. However, in the case of a hybrid zinc-bromine battery, the tank size and power output are interdependent. The advantages of using these hybrid systems are they have greater power density compared to that of nonhybrid flow batteries. The efficiency of hybrid systems is about 75%. The main drawbacks with this system is that bromine is highly toxic to the environment [16].

Capacitors:

The conventional capacitors or capacitors consist of two parallel metal electrodes in which a dielectric medium (such as mica, paper, glass, etc) is sandwiched between the metal plates. On the application of electric field, the charges get polarized and separated in the dielectric medium. This charge separation allows the capacitor to store energy. The capacitance is given by the amount of charge separation stored per unit electric potential which has a unit of farad (F). The capacitance of the capacitor is given by:

$$C = \frac{Q}{V} = \frac{\varepsilon A}{D} \qquad (3)$$

where, amount of charge stored is given by Q, ε is the dielectric medium permittivity, D is the distance between the two electrodes and A is the surface area of electrodes respectively.

The amount of charge stored in the conventional capacitors are in the order of picofarad to millifarad. The primary characteristics of a capacitor are power density and its energy density. To calculate any of these, the density must be calculated with respect to unit mass or unit volume. The energy E stored in a capacitor is directly proportional to its capacitance:

$$E = \frac{1}{2}CV^2 \qquad (4)$$

The amount of energy used per unit time is known as power P. In order to calculate the power of capacitor, it is necessary to take into account that capacitors are often shown as a circuit connected in series with an external load resistance (R). The components of the capacitor such as current collectors, dielectric medium and the electrodes dictates the resistance of the capacitor. This cumulative resistance is known as the equivalent series resistance (ESR). The maximum power P_{max} of the capacitor is limited by its ESR [17], which is evaluated in terms of impedance (R=ESR). This relation is described by the equation.

$$P_{max} = V^2/4 \times ESR \qquad (5)$$

Conventional capacitors have higher power density compared to batteries; however their energy densities are relatively lower. This means batteries has more energy storage capacity than a capacitor, however batteries cannot deliver it as quickly, resulting in a lower power density. Contrarily, capacitors have a lower energy storage capacity. However, the stored electrical energy can be utilized rapidly resulting in an enormous power output. Therefore, capacitors have a high-power density.

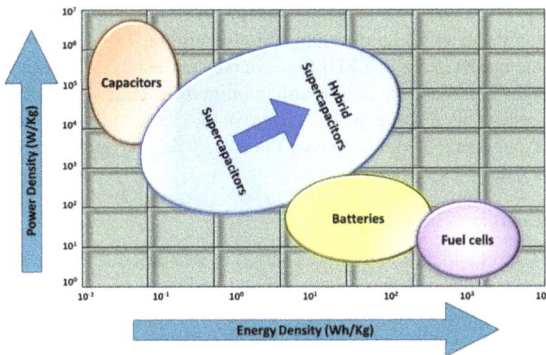

Figure 2. The ragone plot of different energy storage systems.

5.   Supercapacitors

Supercapacitors are often referred to as ultra-capacitors, electrochemical capacitors, or farad capacitors. Which is similar to that of conventional capacitors which have two electrodes separated by a separator soaked in an electrolyte instead of dielectric medium and adhere to some of the same fundamental principles of conventional capacitors, but they use electrodes with greater surface area (A) and thinner dielectric medium (separator) to reduce the separation distance (D) between the electrodes. Therefore, from Eq. 3 and 4, it is evident that this results in an improved capacitance and energy. Supercapacitors can also exhibit high power density like conventional capacitors given that they are maintained with low ESR values. Supercapacitors have many advantages over batteries, such as lower charging time, extended cycle life and superior power density.

Based on the charge storage mechanisms, the supercapacitors are classified into Electric Double-Layer capacitors (EDLCs), pseudocapacitors and hybrid supercapacitors. Fig. 3a represents the different types of supercapacitors and Fig. 3b shows their charge storage mechanisms.

5.1   Electric double layer capacitors (EDLCs)

EDLCs are made up of carbonaceous compounds such as activated carbon, graphene, CNTs, carbon aerogels, etc. which store charges electrostatically which is a non-faradic process, where no redox reactions occur in between both electrodes and electrolyte. The fundamental mechanism behind the energy storage in EDLCs is due to the formation of electrical double layer near electrode surface. When a voltage potential is applied between the two electrodes, the positive charges in the electrolyte accumulate at the negative electrode and vice versa. While accumulating towards the electrodes the ions diffuse through the separator and the pores of the electrodes. In order to prevent the ions pairing at the electrode, a double layer of charges is created. The high specific surface area, with decreased electrode separation distances and the formation of double layer increases the energy density of the EDLCs [18,19].

The energy storing mechanism of EDLCs enables rapid energy capture, release, and improved power performance. The absence of chemical reaction via a non-Faradic mechanism prevents the swell up observed in batteries while charging and discharging. The distinguishing characteristics between EDLCs and batteries are that EDLCs have outstanding cycle stability of around millions of cycles, whereas the batteries can last for only thousands of cycles at best. Still, EDLCs have a restricted energy density as a result of the electrostatic charge storing mechanism. This is the primary reason for the researchers to increase the energy density and to expand the temperature range beyond the capabilities of batteries.

5.2   Pseudocapacitors

In contrast to EDLCs, pseudocapacitors store charges by a faradic surface redox mechanism, where charge is transferred rapidly between electrolyte and surface of electrode. In other words, redox reactions (oxidation and reduction) takes place under the application of potential in pseudocapacitor material. This process entails the movement of charge across the double layer, leading to the flow of faradic current through the supercapacitor cell. Pseudocapacitors achieve high specific capacitance and energy density when compared with carbon-based EDLCs which is due to the faradic reaction. Due to the faradic reaction small oxidation and reduction peaks are visible in cyclic voltammetry, whereas for EDLC type materials the cyclic voltammogram shows a rectangular shape. Transition metal oxides, chalcogenides, conducting polymers such as polypyrrole, PEDOT-PSS, polythiophene, etc. are some examples of electrode material for pseudocapacitors. The faradic behavior involves oxidation and reduction reactions similar to that of batteries; so, they suffer low power density and cycle stability compared to EDLCs. Typically faradic electrode material's specific capacitance lies around 300 to 1000 F.g^{-1}, while EDLCs specific capacitance lies between 100 to 300 F.g^{-1} [20,21].

5.3   Hybrid supercapacitors

EDLC has favorable power performance and cyclic stability, while pseudocapacitors exhibit higher capacitance. The hybrid capacitors combine the charge storage capability of pseudocapacitors and cycle stability, power density, of EDLCs in the same package. With suitable electrode combination, specific capacitance, cell voltage, cyclic stability, power and energy density can be enhanced greatly.

Many have examined different combinations of negative and positive electrodes in organic, aqueous, and non-aqueous electrolytes. In general, the use of faradic electrodes leads to higher capacitance but at the expense of cyclic stability. Presently, researchers are primarily studying three different categories of hybrid supercapacitors, which are classified based on the electrode arrangements such as asymmetric, composite and battery-type.

5.3.1   Asymmetric supercapacitors

Asymmetric hybrid supercapacitors combine non faradic EDLC based material as one electrode and faradic pseudocapacitor behavior material as another electrode. The commonly used configuration is carbonaceous electrode serves as negative electrode and positive electrode is made up of metal oxides, chalcogenides or conducting polymers. In general, pseudocapacitive electrodes have higher energy density and lower resistance when compared to carbon-based electrodes. Nevertheless, pseudocapacitor have lower working potential and low cycling stability. However, Asymmetric supercapacitors exhibit better cycle stability as compared to symmetric pseudocapacitor.

5.3.2   Composite supercapacitors

In composite-type supercapacitors both pseudocapacitive and EDLC behavior materials are combined together to form a single electrode. These components work together in a single electrode to contribute to both the electrostatic and faradaic charge storage processes. The carbonaceous materials provide high specific area, which facilitates improved contact between the pseudocapacitive material, and the electrolyte utilized. The inclusion of pseudo capacitive material in composite material enhances its specific capacitance because of the faradaic reactions. So, the synergistic effect between the composite enables superior energy density, power density, higher specific capacitance, good cycling stability and better coulombic efficiency.

5.3.3   Battery-type Supercapacitors

This type of supercapacitor is similar to that of asymmetric and composite supercapacitors, which uses battery behavior electrode material and supercapacitor materials. This is done to exploit the advantages like high energy density of batteries and high-power density of capacitors. Lithiated and hydroxide-based materials are currently being explored by researchers from all over the world for battery type supercapacitors.

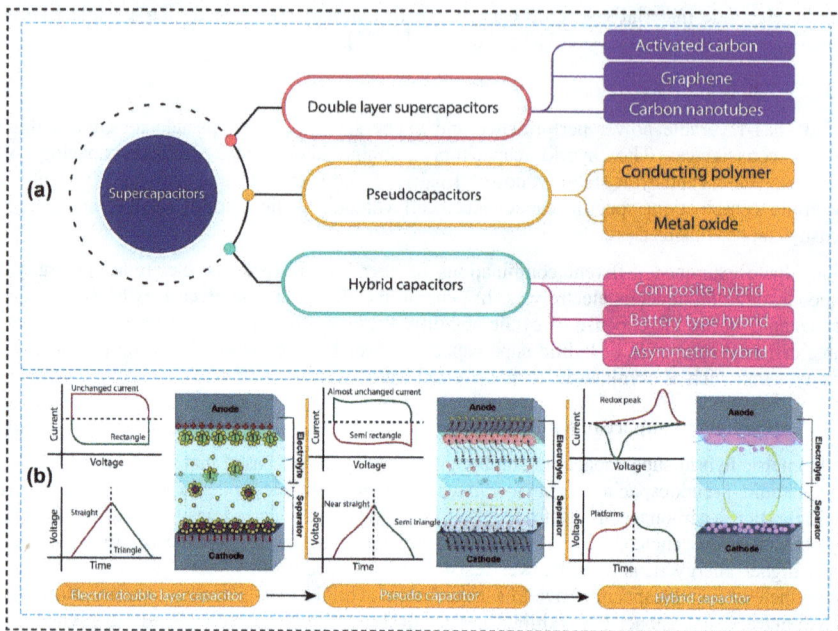

Figure 3. a) types of supercapacitors. b) charge storage mechanism of different SCs. "Reprinted from journal of Energy Storage, volume 81,110430, Ahmad. et. al. 2024, Recent developments in transition metal oxide-based electrode composites for supercapacitor applications, with permission from Elsevier,2024 [22].

5.4   Electrolytes for supercapacitors

Electrolytes are crucial components of supercapacitor designs. They have a significant impact on the operational potential window, power density, energy density and also act as separators in SCs devices [23,24]. In order to meet the demands of supercapacitor applications, the electrolyte must possess the following characteristics. These characteristics encompass low cost, excellent ionic conductivity resulting in efficient power delivery., high concentrations of ions, reduced ion recombination, low toxicity, stable in open atmosphere with good working temperature range and consistent electrochemical performance. Typically, supercapacitors utilize Liquid based electrolytes (ionic, aqueous and organic liquids) and polymer-based electrolytes[24,25].

5.4.1 Liquid electrolytes

Aqueous electrolytes exhibit high dissociation constant so, they are strong electrolytes possessing a larger ion concentration, lower ion-pairing, smaller solvated ions, high dielectric constant and high conductivity of around 1 S.cm^{-1}. These properties result in exhibiting good specific capacitance, energy and power density compared to that of organic electrolyte [24,25]. The common aqueous electrolytes are KOH, H_2SO_4, Na_2SO_4 and so on.

Organic electrolytes have an advantage of providing higher working potential of around 2.5 to 3.0V. The common organic electrolyte is the solution containing $(C_2H_5)_4NBF_4$ in propylene carbonate or acetonitrile. The power of these electrolytes is low compared to aqueous electrolytes due to their low conductivity. Other than that, organic electrolytes have several drawbacks, such as the need for sophisticated purification procedures, inflammable and toxic in nature [26].

Liquid electrolytes have advantages such as high specific capacitance, power and energy density. But these electrolytes are limited due to, corrosion, self-discharge and leakage, Hence, the polymer-based electrolytes are proposed for super capacitor and rechargeable batteries owing to its ionic conductivity and good mechanical stability [21].

5.4.2 Polymer-based electrolytes

Polymer based electrolytes are mechanically strong, have ionic conductivity greater than or equal to 10^{-4} S.cm^{-1}, compatible with most of the electrodes and has better thermal, chemical, and electrochemical stabilities. The commonly used polymer-based electrodes are made by ionic salts such as $LiClO_4$, Na-triflate, $LiBF_4$ and so on dissolved in polymer hosts such as polyethylene Oxide (PEO), Polyphenylene Oxide (PPO), etc. Plasticized polymer electrolytes use liquid plasticizers such as ethylene carbonate or propylene carbonate or polyethylene-glycol is used to dissolve polar salts. Solvent-swollen polymers electrolyte uses Polyvinylpyrrolidone (PVP), Polyvinyl alcohol (PVA) or other polymer matrix in which ionic electrolytes such as H_3PO_4, H_2SO_4 are mixed to form a solid electrolyte. Gel-polymer electrolytes have gained huge attention in recent years, in which liquid or organic electrolytes are mixed with host polymer such as PEO, Polyvinylidene fluoride (PVDF), Polyvinylidene fluoride-co-hexafluoropropylene (PVDF-HFP), Polyether Ether Ketone (PEEK), Polyaryl ether ketone (PAEK), Polyacrylic acid (PAA), Polyacrylonitrile (PAN), etc and dried so that the quasi solid electrolyte exhibit a ionic conductivity similar to that of liquid electrolyte with values reaching upto 10^{-3} S.cm^{-1}, with better stability compared to liquid electrolyte [27,28].

5.4.3 Redox electrolytes

Recently, an entirely new group of electrolytes has been developed by incorporating electroactive substances or using redox mediators or additives in traditional electrolytes to enhance the efficiency of supercapacitors. The redox additive liquid electrolytes are directly added to liquid electrolyte systems which are referred to as redox additive electrolytes. E.g. addition of hydroquinone in H_2SO_4 electrolyte and addition of potassium ferricyanide in KOH electrolyte. These redox additives directly take part in faradaic reactions and improve the electron transfer due to their pseudocapacitve nature. In the case of redox-active electrolytes the entire electrolyte solution is made up of redox species without a supporting liquid electrolyte such as H_2SO_4, KOH and so on. Redox additive-polymer electrolyte is prepared similar to that of polymer electrolyte. But with some addition of redox electrolyte the polymers [21].

6.   MXenes

MXenes are a novel class of two-dimensional materials that were first discovered in 2011 by etching MAX phase. The first discovery of MXene occurred when the Aluminum layer of MAX phase was etched, resulting in the formation of 2D flakes with the general chemical formula of $M_{n+1}X_nT_x$ (n=1,2,3) where M is early transition metals, X is carbon or/and nitrogen and T_x stands for the surface terminations (such as -F, -OH, -O). Therefore, MXenes are class of two-dimensional crystalline materials that have infinite lateral dimension with a negligible thickness. The characteristics of MXenes is also determined by the surface terminated atoms [29]. Some examples for MAX phases are Ti_2AlC, Ti_3AlC_2, Ti_2AlN, Ti_4AlN_3, V_2AlC, V_4AlC_3, Nb_2AlC, Nb_4AlC_3, Zr_3AlC_2, Ta_4AlC_3 and so on. These MAX phases can be exfoliated into Ti_2C, Ti_3C_2, Ti_4N_3, V_2C, V_4C_3, Nb_2C, Nb_4C_3, Zr_3C_2, Ta_4C_3 respectively. The different MXenes structures and their chemical formula can be seen in Fig. 4. Apart from single transition metal MAX phase and MXene, combination of two or more metal can also exist as MAX phase and MXene.

After the discovery of Ti_3C_2 MXene, many scientific groups are working with all sorts of combinations to be used in the fields of energy storage, energy conversion, environmental application, optics, electromagnetic shielding, biotechnology and so on.

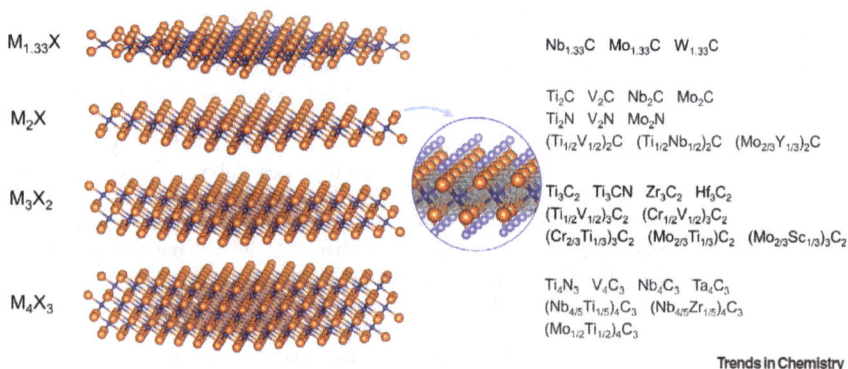

Figure 4. Structural and general formula of experimentally synthesized MXenes. "Reprinted from Trends in Chemistry, volume 1, Issue 7, Verger. et. al. 2019 MXenes: An Introduction of their synthesis, select properties and applications, with permission from Elsevier, 2024 [30].

6.1   Synthesis of MXenes

Layered MAX phase is the precursor for preparation of MXenes which has a general formula of $M_{n+1}AX_n$ (n=1,2,3) where M stands for an early transition metal, A is Aluminum and X is carbon and/or nitrogen. Generally, MAX phase is synthesized by physical route aid of high temperature furnaces. The well-established technique for synthesizing of MXene is by the wet chemical etching method of atomic layers from multi-layer MAX phase, however other methods are also employed to prepare MXene in more environmentally benign ways.

6.1.1 HF etching

In typical HF etching route, MAX phase (e.g. Ti_3AlC_2) is immersed in aqueous solution of hydrofluoric acid. The HF selectively etches the Al atoms in the MAX phase leaving behind two dimensional MXene. Ti_3C_2 is formed by the following reaction:

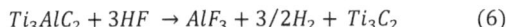

$$Ti_3AlC_2 + 3HF \rightarrow AlF_3 + 3/2H_2 + Ti_3C_2 \qquad (6)$$

$$Ti_3C_2 + 2H_2O \rightarrow Ti_3C_2(OH)_2 + H_2 \qquad (7)$$

$$Ti_3C_2 + 2HF \rightarrow Ti_3C_2F_2 + H_2 \qquad (8)$$

Eq. 6 describes the process in which aluminum atoms are separated from the Ti_3AlC_2 MAX phase to create Ti_3C_2. Following the etching process, the Ti atoms that are exposed will serve as the sites for chemical reactions. These atoms readily react with HF and water as denoted in Eq. 7 and Eq. 8 to form $Ti_3C_2(F)_2$ and $Ti_3C_2(OH)_2$ respectively. The surface termination of hydroxide and fluoride occur simultaneously. The mixture solution is washed with water and ethanol to remove all the impurities to obtain hydroxide and fluoride terminated Ti_3C_2 MXene [31].

Since HF is a hazardous and toxic chemical which is highly corrosive to environment and living beings. Alternatively, a solution mixture of LiF and HCl is subsequently created as a less harsh etchant. As shown in Eq. 9 the concentration of the HF may be controlled by producing it in situ from LiF and HCl. This in-situ produced HF acts as an etchant for the MAX phase. Remarkably, MXene created via *mild etching* technique often exhibits fewer defects and greater planar size compared to MXene produced using HF etching [32].

$$LiF + HCl \rightarrow HF + LiCl \qquad (9)$$

Employing NH_4F is similar to that of using LiF, where NH_4F reacts in the hydrothermal kettle with the Ti_3AlC_2 to form Ti_3C_2 at a certain applied temperature[33]. The below equation given the reaction mechanism:

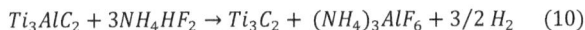

$$Ti_3AlC_2 + 3NH_4HF_2 \rightarrow Ti_3C_2 + (NH_4)_3AlF_6 + 3/2\, H_2 \qquad (10)$$

While the improved LiF and HCl etching process does not directly use HF, but the underlying risk remains the same. This issue could be resolved by the development of different techniques which circumvent the use of HF. The following approaches are explained below.

6.1.2 Urea glass route

In 2015, Mo_2C and Mo_2N MXenes were synthesized from $MoCl_5$ metal precursor via urea glass method. In this method ethanol is introduced to the precursor, resulting in the formation of Mo-orthoesters. Subsequently, Urea was introduced into the solution and the resulting mixture was agitated until complete dissolution of urea. The gel-like material was subjected to heating at a temperature of 800°C in the presence of N_2 gas flow, and subsequently calcinated afterward. The obtained silvery-black powder was analyzed with X-Ray Diffraction (XRD) which had no impurity peaks of MoO_x and other phases were seen. This indicates this method can be employed to obtain high purity MXenes without the aid of toxic HF[34].

6.1.3 Molten salt etching

This technique was first used in 2016 to synthesize MXenes. In this procedure Ti_4AlN_3 MAX phase powder was blended with fluoride salt in a mass ratio of 1:1. The mixture was then heated to a temperature of 550°C for 30 minutes. Finally, the fluoride salts reacted with Al to form different phases of aluminum fluoride but no titanium containing fluorides were found. Which validates Aluminum etching process selective in nature. The Aluminum containing fluorides were dissolved using a solution of dilute sulfuric acid. The resultant impurities were then removed by centrifugating and washing steps. Then the $Ti_4N_3T_x$ MXene were further exfoliated by mixing the MXene powder with tetrabutylammonium hydroxide, which was then rinsed off with DI water with the aid of centrifuge. Further probe sonication and centrifugation were used to obtain smaller and exfoliated MXene sheets which were filtered and used. The resulting MXene exhibited a higher number of atomic defects compared to the MXene synthesized by HF etching [35].

6.1.4   Chemical vapor deposition

The chemical vapor deposition (CVD) technique to produce MXene was first used in 2015. In this technique methane is used as carbon source and a copper foil on top of molybdenum acts as substrate. At temperatures exceeding 1085°C, the chemicals on the tip of substrate decomposes and forms a layer on the substrate from the vapor phase of the materials. At high temperature Copper foil melts and forms a Cu-Mo alloy at the liquid Cu-Mo interface. As a result, Mo atoms diffuse through the Cu liquid surface, which forms Mo_2C crystals via the reaction of carbon atoms generated by the decomposition of methane. However, the resultant product's structure is similar to that of MXene. It is in fact a two-dimensional metal carbide with greater surface area compared to the previously obtained nanosheets, which only reached up to 10 µm. But the CVD grown Mo_2C crystals had a thickness of few nanometers and lateral dimensions exceeding 100 µm. The thickness of sheets can be controlled by adjusting the methane concentration. Also, the MXene-like material obtained is defect-free when compared to molten salt synthesizing route [36].

6.1.5   Hydrothermal synthesis

In 2018 the hydrothermal method is used to prepare $Ti_3C_2T_x$ MXene from the max phase of Ti_3AlC_2 with the aid of sodium hydroxide. This method involves the hydroxide anions $(OH)^-$ attacking the layers of aluminum atoms, leading to the formation of aluminum hydroxide. The exposed Ti atoms are surface terminated by -OH or -O. The aluminum hydroxide confined by Ti layers does not allow O or OH to react with Ti. This issue is solved by repeating the process with different concentration of NaOH and different hydrothermal temperatures. The MXenes synthesized using the hydrothermal method exhibit a higher number of -OH and -O terminations compared to those formed using HF etching. This substantial increase in terminations greatly improves the performance as supercapacitors [37]

6.1.6   Electrochemical synthesis

In 2018, an electrochemical approach was first suggested for the delamination of Ti_3C_2 in a binary aqueous electrolyte, without the use of fluorine. A- two electrode system was built using bulk MAX phase of Ti_3AlC_2 which serves as both anode and cathode. The etching underwent only in anode, resulting in the production of $Ti_3C_2T_x$. In order to prevent only the surface etching and facilitate the diffusion of electrolyte into the deeper layers of anode, A mixture of 0.2 M tetramethylammonium hydroxide (TMAOH) and 1M of ammonium chloride with a pH of 9 was maintained. By applying low potential of 5V, the bulk anode underwent a progressive delamination. The settled and suspended particles were pulverized and mixed with a solution containing 25% w/w of TMAOH to exfoliate the MXene sheets [38,39]. The resulting MXenes

Emerging Materials for Next Frontier Energy and Environment Applications   Materials Research Forum LLC
Materials Research Foundations 170 (2024) 117-144   https://doi.org/10.21741/9781644903292-6

demonstrated electrical conductivity similar to that of MXenes obtained from HF and HF/LiF etching procedure. Nevertheless, the electrochemical method seems to be the most favorable in terms of total etching yield, which is around 60% of bulk MAX phase to MXenes.

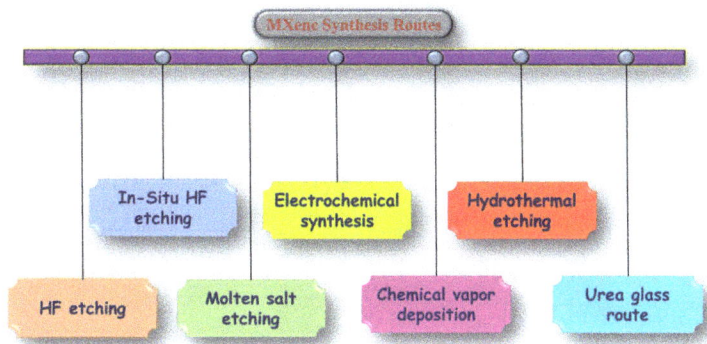

Figure 5. Different synthesis techniques of MXenes.

6.3   MXene properties

MXenes mechanical properties directly depend on the surface terminations and the nature of the material used in MXene. MXenes possess high stiffness due to the bonding between the Ti-O, which is the strongest when compared with Ti-F and Ti-OH. Applications demanding high structural stability and mechanical strength oxygen terminated MXenes are the most suitable candidates [40]. By evaluating the Young's modulus, it was shown that the stiffness of the carbides increases with the increase in transition metal mass. The stiffness and strength of MXene is lower than graphene. Although this fact, Molecular dynamics simulations have predicted significantly high level of stiffness for Ti_2C, Ti_3C_2 and Ti_4C_3 MXenes. Also, thicker MXenes exhibit higher bending rigidity values compared with MoS_2 and graphene [41].

Most of the MXenes and their different surface terminations have almost metallic behavior in terms of conductivity. Among the different MXenes, the first discovered $Ti_3C_2T_x$ is still the most conductive one [42]. One general approach to alter MXenes conductivity is by altering the cations or intercalating large organic molecules which changes the resistance in an order of magnitude. Also, using reinforcing materials such as chitosan to manipulate the electrical conductivity, by increasing the chitosan loading the resistance of the composite also increases. MXenes with -OH termination possess nearly unbound electron states, which are located parallel and outside the surface atoms. These states are found near the highest positive charge concentration, which serves as an ideal pathway for electron transport. Similarly, the experiment with $Ti_{n+1}C_nT_x$ MXenes, revealed that -OH terminated MXenes had highest conductivity when compared with -F termination, which had good conductivity with respect to -O terminated MXenes [43].

The general MXenes such as Ti_3C_2, are non-magnetic in nature, however introduction of magnetic cations like Mn ($Ti_2MnC_2T_x$) in the system can alter the non-magnetic material to ferromagnetic in nature even at ground states irrespective of the surface terminations. But some intrinsically magnetic natured MXenes are predicted which are Cr_2C, Ti_2N (ferromagnetic), Cr_2N and Mn_2C (anti-ferromagnetic). Nitrogen based MXenes have an additional electron in their unit cell when compared to carbon based MXenes. The theoretical model predicted five nitrogen-based MXenes which are ferromagnetic in nature at their ground states which are Cr_2NO_2, Ti_2NO_2, $Mn_2N(OH)_2$, Mn_2NO_2, Mn_2NF_2. Creating intrinsic point defect in MXenes also create magnetic nature in MXene due to unpaired electrons in the spin-split d orbitals [44,45].

MXenes have low surface area compared to that of carbon-based materials, for instance, carbon black has a Specific surface Area (SSA) of around 900 m^2g^{-1} while carbon nanotubes have a SSA of around 100-1000 m^2g^{-1} and a single sheet of graphene has a theoretical SSA of 2630 m^2g^{-1}. Whereas experimentally derived Ti_3C_2 MXene shows a SSA of 66 m^2g^{-1}, V_2CT_x exhibits a SSA value of 19 m^2g^{-1}. While theoretical $Ti_3C_2T_x$ can reach upto 496m^2g^{-1}. Even with low surface area when seen in terms of volumetric capacity, functionalized graphene has shown values of upto 200 F/g. While the MXene achieved a capacity of 320 F/g even in its early days [41,46].

7. Transition metal oxides for supercapacitors

TMOs are chemically bonded to oxygen atoms which consist of both pseudocapacitance and battery-like behavior material. The incomplete d-subshell of these TMOs allow variations in oxidation state, which in turn contribute to their electronic, structural, and electrochemical characteristics. Reducing the size of bulk materials to the nanoscale regime allows for changes in surface properties, which leads to increased specific surface area. This facilitates the efficient usage of electrode materials. The TMOs exhibit excellent conductivity, negligible structural changes, multiple oxidation state in the working potential and demonstrate fast cation, electron or proton transfer between electrode and electrolyte during electrochemical reactions. In 1972, RuO_2 was the first material investigated as a pseudocapacitive material because of its multiple oxidation states and fast reversible redox reactions[47]. This material has many advantageous characteristics, such as high energy density, high power density, exceptional electronic conductivity, and favorable thermal stability which makes it a viable candidate for energy storage electrode material. During the faradaic redox reactions, the ions in the electrolyte deeply infiltrate the electrode's surface and then release reversibly through empty voids into the electrolyte. Amorphous RuO_2 has been discovered to possess a higher specific capacitance compared to that of crystalline material due to the larger diffusion of protons into amorphous solids than that of crystalline solids. The mechanism involves the storage of charges in an acidic electrolyte medium by the adsorption of protons onto the surface of RuO_2. This process is accompanied by fast electron transfer, resulting in a shift in the oxidation state from Ru(II) to Ru(IV) [48].

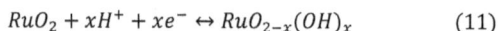

$$RuO_2 + xH^+ + xe^- \leftrightarrow RuO_{2-x}(OH)_x \qquad (11)$$

The anhydrous form of ruthenium oxide (RuO_2) has a lower ionic conductivity compared to hydrous ruthenium oxide ($RuO_2.H_2O$) because of the diffusion of H^+ cation between OH^- and H_2O entities. The electrochemical characteristics of the material are greatly dictated by the material's surface area, active sites, electronic conductivity, multiple oxidation states and structural stability. A highly porous nanomaterial facilitates efficient and rapid transport of electrolyte ions during the charging and discharging processes, resulting in better energy storage performance of supercapacitors. RuO_2, being a noble metal oxide, exhibits one of best performance for supercapacitors. However, the main drawback is its exorbitant cost. The need of the hour is to explore non-noble metal-based electrode material and to exploit its properties to achieve a similar performance to that of noble metals. In that scenario, Mn, Co, Ni, Cu, Zn, W, Mo, Fe, V, and so on based oxides are being explored for their electrochemical capabilities. These TMOs can be synthesized in numerous methods from a simple co-precipitation method to sophisticated atomic layer deposition. Depending upon the desirable material properties and morphology, suitable techniques can be employed to achieve the same. Fig. 6 shows the top-down and bottom-up approaches with some examples through which the TMOs can be prepared. Also compositing and creating heterojunction between various metal oxides and 2D materials is a feasible way to achieve outstanding supercapacitor performance in cost-effective manner.

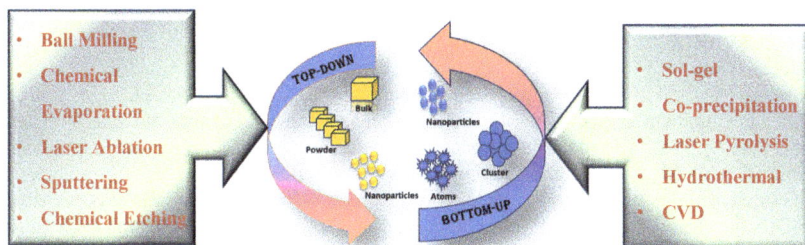

Figure 6. Different techniques to synthesize TMOs.

8.   MXene-transition metal oxide nanocomposite for supercapacitors

The synergistic effect between the Transition metal oxides and the highly conductive MXene for the application of supercapacitors will be discussed in this section.

Iron oxide nanorods were coupled with Ti_3C_2 MXenes were used as supercapacitors electrode material by Li et. al. The Ti_3C_2 MXene were synthesized by in situ HF production using LiF salts and HCl. The Fe_2O_3 nanorods were directly grown on carbon cloth with the aid of hydrothermal method. Where, FOOH were grown on carbon cloth after hydrothermal treatment. Then the samples were annealed for 2 hours at 450°C in nitrogen atmosphere. To prepare MXene-iron oxide composite on carbon cloth. The as-prepared MXene nano sheets were dispersed in DI water by sonication to make a homogenous solution. Further, the previously prepared Fe_2O_3 nanorods on carbon cloth was completely immersed into the MXene solution. The MXene soaked carbon cloth is dried at 80°C. The dipping and drying process was repeated several times to reach the desired mass loading. This MXene-Fe_2O_3 composite on carbon cloth is used as negative electrode for flexible solid-state asymmetric supercapacitor device. Whereas the positive electrode is made up of MnO_2 on activated carbon cloth. The activation of carbon cloth is done by soaking carbon cloth in solution containing dopamine hydrochloride and Tris-HCl for 2 hours. The pH was maintained at 8.5 with the addition of 1 M of NaOH solution and stirred continuously for 24 hours at 60°C. Then the carbon cloth was cleaned and annealed at 1000°C in N_2 atmosphere for 1 hour. The activated carbon cloth was immersed in 10 mM potassium permanganate solution for 6 hours at 90°C to grow MnO_2 nanoparticles on carbon cloth. The three-electrode measurement consists of saturated Ag/AgCl electrode as reference electrode, Pt plate as counter electrode, the previously prepared composites on carbon cloth as working electrode and 5 M LiCl acts as aqueous electrolyte. The areal capacitance of individual MXene and Fe_2O_3 on carbon cloth was 136.6 $mF.cm^{-2}$ and 206.38 $mF.cm^{-2}$ at a current density of 1 $mA.cm^{-2}$. The composite prepared with MXene and Fe_2O_3 showed an exceptional capacitance of 725 $mF.cm^{-2}$ at 1 $mA.cm^{-2}$. The electrodes exhibited an excellent retention of 88.6% after 10,000 cycles. While the pristine Fe_2O_3 electrode had a retention of only 42%. The prepared MnO_2 electrode demonstrated an outstanding areal capacitance of 1065 $mF.cm^{-2}$ at 1 $mA.cm^{-2}$. The prepared asymmetric device had a potential window of 1.8V, which showed a specific capacitance of 143.4 $mF.cm^{-2}$ at 1 $mA.cm^{-2}$. The stability of 78.7% after 5000 cycles was demonstrated by the asymmetric device. The energy density of the device is 1.61 $mWh.cm^{-3}$ at an impressive power density of 22.6 $mW.cm^{-2}$. The fabricated device lit up a red LED for 30 seconds [49].

Tian et.al. decorated Mn_2O_3 and MnO nanoparticles on two dimensional MXene for supercapacitor electrodes. Firstly, Ti_3C_2 MXenes were delaminated from Ti_3AlC_2 MAX phase using LiF and HCl to etch the aluminum atoms in the MAX phase while stirring for 24 hours at a constant heating temperature of 35°C. The obtained samples were washed thoroughly to remove impurities and delamination of MXene is carried out by using ultrasonication for 1 hour. The delaminated MXenes were collected and stored. MnO_x was grown on MXene by dispersing layered MXenes into DI water and manganese nitrate solution is added to the MXene dispersion with constant stirring for 24 hours and sonicated for 1 hour. The suspension is filtered with membrane and the obtained film is carefully dried and annealed at nitrogen atmosphere for 2 hours at 300°C. The three electrode studies were carried out with saturated calomel electrode and platinum sheet as reference and counter electrode respectively, 1M $LiSO_4$ is used as the electrolyte. The obtained film is used as a working electrode. The volumetric capacitance of Manganese oxide at Ti_3C_2 electrode is higher than that of pure Ti_3C_2 electrodes which is 602 $F.cm^{-3}$ and 445.2 $F.cm^{-3}$ respectively at 2 $mV.s^{-1}$. Two electrode studies were carried out with the same MnO_x MXene composite film for both positive and negative electrode. The symmetric device exhibited a capacitance of 392.9 $F.cm^{-3}$ at 2 $mV.s^{-1}$ scan rate, the energy and power density of the device are 13.64 $mWh.cm^{-3}$ at 2 $mV.s^{-1}$ and 3755.61 $mW.cm^{-3}$ at 100 $mV.s^{-1}$. The device exhibited an capacitance retention of 89.9% after 10,000 cycles [50].

Chavan et.al synthesized Ti_3C_2 MXene via HF etching route and nickel oxide by coprecipitation method followed by annealing. The nanocomposite was synthesized by different weight percentage of NiO (5, 10, 15, 20 wt.%) were dispersed in ethanol along with MXene and ultrasonicated under a constant temperature of 60°C. After evaporation of ethanol, the particles were collected and annealed for 300°C for 1 hour under nitrogen atmosphere. The slurry was made with acetylene black, PVDF and NMP. The slurry was screen printed into flexible stainless-steel mesh which acts as working electrode. In three-electrode system, graphite rod and saturated Ag/AgCl were used as counter and reference electrodes, the electrolyte used for ion transportation was 1 M KOH. The 15 wt% NiO@MXene exhibited the highest specific capacitance of 1542 $F.g^{-1}$ at a current density of 6 $mA.cm^{-2}$. This electrode material is used to make an all-solid-state asymmetric supercapacitor where CuO was employed as negative electrode and gel based PVA-KOH electrolyte was utilized. The fabricated device exhibited a specific capacitance of 73.3 $F.g^{-1}$ at a current density of 10 $mA.cm^{-1}$, power density of 3.3 $KW.kg^{-1}$ even at a energy density of 10.5 $Wh.kg^{-1}$ and the capacitance retention of the device was 90.6% even after 5000 cycles [51].

Vigneshwaran et. al. nanocomposited hydrothermally prepared $Ni-CoWO_4$ with HF etched Ti_3C_2 MXene were processed using sonication and annealing technique. Carbon cloth was used as working substrate, saturated Ag/AgCl electrode as reference and platinum wire was utilized as counter electrode for three-electrode measurements. The Acidic 1 M H_2SO_4 was used as electrolyte. The composite material exhibited a good capacitance of 582 $F.g^{-1}$ at a current density of 1 $A.g^{-1}$. In three electrode system the capacitance retention exhibited was 93.5% for 10,000 cycles. Symmetric supercapacitor coil cell was fabricated with the same composite material which acts as both positive and negative electrode. The designed device revealed an energy density of 85.7 $Wh.kg^{-1}$ even at a power density of 850 $W.kg^{-1}$. Also, 96.5% of initial capacitance was retained by the device even after 10,000 cycles. With the device charged to 1.7 V, it was able to glow up a red LED light [52].

In one of the works by Xia et. al. embedded $MnCo_2O_4$ nanoparticles in the Ti_3C_2 MXene which was obtained through HF etching process. The $MnCo_2O_4$ on Ti_3C_2 MXene was prepared using a hydrothermal method. The obtained material was coated onto a nickel foam which acts as working electrode, platinum mesh as counter electrode and Hg/HgO electrode was used as reference electrode in three electrode system in which, 1 M KOH served as electrolyte for the system. The electrode exhibited an specific capacitance of 806.7 $F.g^{-1}$ at 1 $A.g^{-1}$ current density. An asymmetric supercapacitor was assembled with activated carbon as negative electrode, The device showed a specific capacitance of 20.94 $F.g^{-1}$ at 1 $A.g^{-1}$, power density of 2.88 $kW.kg^{-1}$ at a energy density of 26.8 $Wh.kg^{-1}$. The device had a potential window of 1.6V, with two asymmetric devices connected in series which light up 4 red LEDs. The device showed a stability of 93.8% even at 5000 cycles [53].

Zheng et. al. fabricated a MoO_{3-x}-Ti_3C_2 MXene free-standing film. At first, Molybdenum powder is dispersed into a solution containing H_2O_2 and stirred for 30 minutes at room temperature and DI water was added to the solution. Then the solution is hydrothermally treated for 1 day at a temperature of 140°C. The obtained precipitate was washed thoroughly, dried and stored. MXene was obtained from etching of Ti_3AlC_2 MAX phase with salts of lithium fluoride and HCl. The etched MXene was washed and ultrasonicated under nitrogen atmosphere for 1 hour. MoO_3/Ti_3C_2 composites are fabricated by slowly mixing MoO_3 solution with MXene solution and agitated vigorously for few minutes. The solution is vacuum filtered via a membrane to obtain composite films. The three-electrode setup consists of activated carbon disc electrode as counter electrode, saturated Ag/AgCl as reference electrode, the free-standing Molybdenum-MXene composite film as working electrode and 5 M LiCl as the electrolyte. The specific capacitance of 631 $F.cm^{-3}$ at a current density of 1$A.g^{-1}$ was exhibited by the films along with a capacity retention of 103.9% after 10,000 cycles. An asymmetric device is fabricated with Nitrogen doped activated carbon electrode as used as cathode with 5 M LiCl electrolyte. The device had a working potential window of 2.1V with a specific capacitance 79.3 $F.cm^{-3}$ at a current density of 1 $A.g^{-1}$ and12.6 $F.cm^{-3}$ even at a higher current density of 20 $A.g^{-1}$. The fabricated device exhibited a capacitance retention 96.3% even after 20,000 cycles [54]. Previously Zheng et. al. prepared the same electrode with different electrolyte. In that work 3M H_2SO_4 was used as electrolyte. Where the electrode revealed a capacitance of 837 $C.g^{-1}$ or 1836 $C.cm^{-3}$ at 1 $A.g^{-1}$, 63.8% of retention was maintained even at a very high current density of 100 $A.g^{-1}$. The asymmetric device fabricated with nitrogen-doped activated carbon showed a specific capacity of 39.5 $C.g^{-1}$ or 49.7 $C.cm^{-3}$ and the device also exhibited a capacity retention of 94.2% after 10,000 cycles at a current density of 10 $A.g^{-1}$. The energy and current densities of the devices are 31.2 $Wh.kg^{-1}$ and 39.2 $Wh.L^{-1}$ at a current density of 0.5 $A.g^{-1}$. At a current density of 50 $A.g^{-1}$ the energy density is 37.5 $Wh.kg^{-1}$ and the power density if 47.1 $kW.L^{-1}$ [55].

Table1: Comparison Table of different TMOs composite with MXene with different synthesizing techniques.

MXene/TMO Composite	Synthesis Method	Current Collector	Electrolyte	Specific Capacitance	Cyclic Stability with cycle number	Ref.
$Ti_3C_2T_x/MnO_2$	Chemical Deposition	Ni Foam	0.5M H_2SO_4	242 F/g @ 1A/g	97% (5000)	[56]
$Ti_3C_2T_x/MnO_2$	Chemical deposition	Carbon Cloth	3M KOH	212F/g @ 1A/g	88% (10000)	[57]
$Ti_3C_2T_x/MnO_2$	Chemical deposition	Ni Foam	1M Na_2SO_4	130.5F/g @ 0.2A/g	90% (1000)	[58]
$Ti_3C_2T_x/MnO_2$	Self-Assembly	Stainless Steel	1M Na_2SO_4	340F/g @ 1A/g	87.6% (2000)	[59]
Ti_3C_2/MnO_2	Hydrothermal	Ni Foam	6M KOH	254F/g @ 0.5A/g	95.5% (5000)	[60]
$Ti_3C_2T_x/NiO$	Hydrothermal	Ni Foam	3M KOH	630.9C/g @ 1A/g	92.9% (5000)	[61]
$Ti_3C_2T_x/NiO/TiO_2$	Hydrothermal	Ni Foam	1M KOH	60.7mAh/g @ 1A/g	70.4% (5000)	[62]
$Ti_3C_2T_x/NiO$	Screen Printing	Stainless Steel	1M KOH	1542F/g @ 6mA/cm^2	90.3% (3000)	[51]
$Ti_3C_2T_x/Ni$-$CoWO_4$	Hydrothermal	Carbon Cloth	1M H_2SO_4	582F/g @ 1 A/g	93.5% (10000)	[52]
$Ti_3C_2T_x/MnFe_2O_4$	Precipitation	Ni Foam	2M KOH	1263F/g @ 1 A/g	96.04% (5000)	[63]
$Ti_3C_2T_x/TiO_2$	Insitu Hydrolysis	Ni Foam	6M KOH	143F/g @ 1A/g	92% (6000)	[64]
$Ti_3C_2T_x/Fe_2O_3$/ rGO	Chemical process	Stainless Steel	5M LiCl	45.8F/g @ 0.2A/g	82.1% (5000)	[65]
Ti_3C_2/WO_3	Hydrothermal	Stainless Steel	0.5M H_2SO_4	566F/g @ 5A/g	92.33% (5000)	[66]
$Ti_3C_2T_x/Fe_2O_3$	Hydrothermal	Carbon Cloth	5M LiCl	725mF/cm^2 @ 1mA/cm^2	88.6% (10000)	[49]
$Ti_3C_2T_x/CoFe_2O_4$	Coprecipitation	ITO coated PET	0.1M KOH	1268.75F/g @ 1A/g	97% (5000)	[67]
$Ti_3C_2T_x/ZnO$	Precipitation	Ni Foam	1M KOH	120F/g @ 2mV/s	85% (10000)	[68]
$Ti_3C_2T_x/MnCo_2O_4$	Hydrothermal	Free standing film	1M KOH	806.67F/g @ 1A/g	77% (3000)	[53]
Ti_3C_2/MoO_3	Hydrothermal	Carbon Cloth	2M KOH	775F/g @ 1A/g	96.4% (6000)	[69]
$Ti_3C_2T_x/MoO_3$	Hydrothermal	Ni Foam	1M KOH	151F/g @ 2mV/s	93.7% (8000)	[70]
Ti_3C_2/MoO_{3-x}	Vacuum Filtration	Free standing film	5M LiCl	631F/cm3 @ 1A/g	103% (10000)	[54]
$Ti_3C_2T_x/CoO_x$-NiO	ALD	Ni Foam	6M KOH	1960F/g @ 1A/g	90.2% (8000)	[71]
$Ti_3C_2T_x/rGO@NiCoO_2$	Solvothermal	Carbon Cloth	2M KOH	1662.5F/g @ 0.5A/g	77.29% (10000)	[72]
$Ti_3C_2T_x/Co_3O_4$	Hydrothermal	Ni Foam	6M KOH	1081F/g @ 0.5A/g	-	[73]

Nb$_2$CT$_x$/Nb$_2$O$_5$/ Carbon	CO$_2$ Oxidation	Cu Foil	1MLiClO$_4$/E C/DMC	530C/g @ 1mV/s	-	[74]
Ti$_3$C$_2$/MnO$_x$	Insitu Wet Chemical	Ni Foam	1M Li$_2$SO$_4$	602F/cm^2 @ 2mV/S	-	[50]
Ti$_3$C$_2$T$_x$/PANI @TiO$_2$	Hydrothermal	Ni Foam	1M KOH	188.3F/g @ 10mV/s	94% (8000)	[75]
Ti$_3$C$_2$T$_x$/WO$_3$	Hydrothermal	Stainless Steel	0.5M H$_2$SO$_4$	297F/g @ 1A/g	73.4% (5000)	[76]
Ti$_3$C$_2$/P-doped RuO$_2$	Chemical Synthesis	Stainless Steel	1M H$_2$SO$_4$	612.72F/g @ 2A/g	97.95% (10000)	[77]
Ti$_3$C$_2$T$_x$/NiMoO$_4$	Hydrothermal	Ni Foam	3M KOH	545.5 F/g @0.5A/g	68% (10000)	[78]
Ti$_3$C$_2$T$_x$/Co$_2$NiO$_4$	Hydrothermal	Ni Foam	3M KOH	719.5 F/g @0.5A/g	83.1% (10000)	[79]

Table2: Comparison table for fabricated supercapacitor device performances based on TMOs/MXene active materials.

Electrode (Positive electrode//Negative electrode)	Electrolyte	Specific Capacitance	Cyclic Stability	Energy Density	Power Density	Ref.
Ni-dMXNC// Ti$_3$C$_2$Tx	1M KOH	92 mAh/cm^3	72.1% (5000)	10.4 mWh/cm^3	0.22 W/cm^3	[62]
NiO@MX//CuO	PVA- KOH	73.3 F/g	90.6% (5000)	10.7 Wh/kg	3333 W/kg	[51]
Ti$_3$C$_2$T$_x$/Ni-CoWO$_4$// Ti$_3$C$_2$T$_x$/Ni-CoWO$_4$	1M H$_2$SO$_4$	248 F/g	96.5% (10000)	85.7 Wh/kg	0.85 kW/g	[52]
Mxene@Fe$_2$O$_3$//MnO$_2$	5M LiCl	143.4F/g	78.7% (5000)	0.74 mWh/cm^3	495.6 mW/cm^3	[49]
Ti$_3$C$_2$T$_x$/MnCo$_2$O$_4$// Activated Carbon	1M KOH	20.94F/g	93.8% (5000)	26.8 Wh/kg	2.88 kW/g	[53]
Ti$_3$C$_2$/Co$_3$O$_4$// PANI@CFP	6M KOH	95.71F/g	83% (8000)	26.06 Wh/kg	700 W/g	[73]
Ti$_3$C$_2$/MnO$_x$//Ti$_3$C$_2$/MnO$_x$	1M Li$_2$SO$_4$	392.9F/cm^3	89.8% (10000)	13.64 mWh/cm^3	3755.61 mW/cm^3	[50]
Ti$_3$C$_2$T$_x$/Co$_2$NiO$_4$// Activated Carbon	PVA- KOH	380F/g	90.4% (3500)	49.74 Wh/kg	2752.21 W/kg	[79]
Ti$_3$C$_2$T$_x$/NiMoO$_4$// rGO	3M KOH	150.1 C/g	72.6% (10000)	33.76 Wh/kg	400.08 W/kg	[78]
Ti$_3$C$_2$/MoO$_{3-x}$ //Ni doped AC	5M LiCl	79.3F/cm^3	96.3% (20000)	48.6Wh/L	24.7kW/L	[54]

Conclusion

TMOs have garnered significant attention because of their abundant availability, cost-effectiveness, eco-friendly nature, and high specific capacitance, rendering them suitable for the use of electrode materials for high performance supercapacitor devices. However, the poor conductivity hinders the performance of TMOS based SCs. To enhance the supercapacitor performance of TMO based electrode material, a viable approach would be to create TMO nanocomposite with highly conductive two-dimensional materials such as MXenes. This will facilitate rapid charge transfer. The combination of TMOs with conductive two-dimensional materials has led to improved electrochemical properties and increased shear flexibility. MXenes are regarded as promising support for TMOs due to their ability to enhance conductivity and improve electrochemical performance. The combination of hybrid TMOs with MXenes offers a wide range of improvements that cannot be attained by using an individual material. For instance, combining MXene with TMOs, results in synergistic effects by which MXene contributes to large electronic transport due to its extraordinary conductivity and increases the specific surface area while maintaining material's structural and chemical stability. On the other hand, the electroactive TMOs prevent agglomeration and restacking of TMOS. Both MXene and TMOs enhance the performance of the device by facilitating rapid ion movement and improving ion accessibility via their interlayer spacing. Therefore, the use of MXene/TMOs nanocomposite enhances the performance of supercapacitors and can be commercialized to satisfy the intermittent power requirements, in the applications of grid regulators, electric vehicles, wearables smart electronics, and so on.

References

[1] R.J. Kuhns, G.H. Shaw, Navigating the Energy Maze, Springer International Publishing, Cham, 2018. https://doi.org/10.1007/978-3-319-22783-2.

[2] S. Chu, Y. Cui, N. Liu, The path towards sustainable energy, Nature Mater 16 (2017) 16–22. https://doi.org/10.1038/nmat4834.

[3] S. Singh, S. Jain, V. Ps, A.K. Tiwari, M.R. Nouni, J.K. Pandey, S. Goel, Hydrogen: A sustainable fuel for future of the transport sector, Renewable and Sustainable Energy Reviews 51 (2015) 623–633. https://doi.org/10.1016/j.rser.2015.06.040.

[4] M. Allen, P. Antwi-Agyei, F. Aragon-Durand, M. Babiker, P. Bertoldi, M. Bind, S. Brown, M. Buckeridge, I. Camilloni, A. Cartwright, W. Cramer, P. Dasgupta, A. Diedhiou, R. Djalante, W. Dong, K.L. Ebi, F. Engelbrecht, S. Fifita, J. Ford, S. Fuß, B. Hayward, J.-C. Hourcade, V. Ginzburg, J. Guiot, C. Handa, Y. Hijioka, S. Humphreys, M. Kainuma, J. Kala, M. Kanninen, H. Kheshgi, S. Kobayashi, E. Kriegler, D. Ley, D. Liverman, N. Mahowald, R. Mechler, S. Mehrotra, Y. Mulugetta, L. Mundaca, P. Newman, C. Okereke, A. Payne, R. Perez, P.F. Pinho, A. Revokatova, K. Riahi, S. Schultz, R. Seferian, S. Seneviratne, L. Steg, A.G. Rogriguez, T. Sugiyama, A. Thonas, M.V. Vilarino, M. Wairiu, R. Warren, G. Zhou, K. Zickfeld, Technical Summary: Global warming of 1.5°C. An IPCC Special Report on the impacts of global warming of 1.5°C above pre-industrial levels and related global greenhouse gas emission pathways, in the context of strengthening the global response to the threat of climate change, sustainable development, and efforts to eradicate poverty, (2019). https://www.ipcc.ch/site/assets/uploads/sites/2/2018/12/SR15_TS_High_Res.pdf.

[5] U. Sohail, E. Pervaiz, M. Ali, R. Khosa, A. Shakoor, U. Abdullah, Role of tungsten carbide (WC) and its hybrids in electrochemical water splitting application- A comprehensive review, FlatChem 35 (2022) 100404. https://doi.org/10.1016/j.flatc.2022.100404.

[6] N.Z. Muradov, T.N. Veziroğlu, "Green" path from fossil-based to hydrogen economy: An overview of carbon-neutral technologies, International Journal of Hydrogen Energy 33 (2008) 6804–6839. https://doi.org/10.1016/j.ijhydene.2008.08.054.

[7] A.H. Fathima, K. Palanisamy, Energy Storage Systems for Energy Management of Renewables in Distributed Generation Systems, in: Energy Management of Distributed Generation Systems, IntechOpen, 2016. https://doi.org/10.5772/62766.

[8] M. Yekini Suberu, M. Wazir Mustafa, N. Bashir, Energy storage systems for renewable energy power sector integration and mitigation of intermittency, Renewable and Sustainable Energy Reviews 35 (2014) 499–514. https://doi.org/10.1016/j.rser.2014.04.009.

[9] A.H. Fathima, K. Palanisamy, 8 - Renewable systems and energy storages for hybrid systems, in: A.H. Fathima, N. Prabaharan, K. Palanisamy, A. Kalam, S. Mekhilef, Jackson.J. Justo (Eds.), Hybrid-Renewable Energy Systems in Microgrids, Woodhead Publishing, 2018: pp. 147–164. https://doi.org/10.1016/B978-0-08-102493-5.00008-X.

[10] M.E. Amiryar, K.R. Pullen, A Review of Flywheel Energy Storage System Technologies and Their Applications, Applied Sciences 7 (2017) 286. https://doi.org/10.3390/app7030286.

[11] M. Javaheri, A. Shafiei Ghazani, Energy and exergy analysis of a novel advanced adiabatic compressed air energy storage hybridized with reverse osmosis system, Journal of Energy Storage 73 (2023) 109250. https://doi.org/10.1016/j.est.2023.109250.

[12] C. de M. Altea, J.I. Yanagihara, Energy, exergy and environmental impacts analyses of Pumped Hydro Storage (PHS) and Hydrogen (H2) energy storage processes, Journal of Energy Storage 76 (2024) 109713. https://doi.org/10.1016/j.est.2023.109713.

[13] R.J. Brodd, SECONDARY BATTERIES | Overview, in: J. Garche (Ed.), Encyclopedia of Electrochemical Power Sources, Elsevier, Amsterdam, 2009: pp. 254–261. https://doi.org/10.1016/B978-044452745-5.00125-8.

[14] N. Vangapally, T.R. Penki, Y. Elias, S. Muduli, S. Maddukuri, S. Luski, D. Aurbach, S.K. Martha, Lead-acid batteries and lead–carbon hybrid systems: A review, Journal of Power Sources 579 (2023) 233312. https://doi.org/10.1016/j.jpowsour.2023.233312.

[15] J. Conzen, S. Lakshmipathy, A. Kapahi, S. Kraft, M. DiDomizio, Lithium ion battery energy storage systems (BESS) hazards, Journal of Loss Prevention in the Process Industries 81 (2023) 104932. https://doi.org/10.1016/j.jlp.2022.104932.

[16] J. Ferrari, Chapter 3 - Energy storage and conversion, in: J. Ferrari (Ed.), Electric Utility Resource Planning, Elsevier, 2021: pp. 73–107. https://doi.org/10.1016/B978-0-12-819873-5.00003-4.

[17] M.A. Scibioh, B. Viswanathan, Chapter 2 - Fundamentals and energy storage mechanisms—overview, in: M.A. Scibioh, B. Viswanathan (Eds.), Materials for Supercapacitor Applications, Elsevier, 2020: pp. 15–33. https://doi.org/10.1016/B978-0-12-819858-2.00002-0.

[18] M.V. Kiamahalleh, S.H.S. Zein, G. Najafpour, S.A. Sata, S. Buniran, Multiwalled carbon nanotubes based nanocomposites for supercapacitors: a review of electrode materials, NANO 07 (2012) 1230002. https://doi.org/10.1142/S1793292012300022.

[19] A. Mani, K.Z. Kamali, A. Pandikumar, Y.S. Lim, H.N. Lim, N.M. Huang, Graphene-Polypyrrole Nanocomposite: An Ideal Electroactive Material for High Performance Supercapacitors, in: Graphene Materials, John Wiley & Sons, Ltd, 2015: pp. 225–244. https://doi.org/10.1002/9781119131816.ch7.

[20]   A. Afif, S.M. Rahman, A. Tasfiah Azad, J. Zaini, M.A. Islan, A.K. Azad, Advanced materials and technologies for hybrid supercapacitors for energy storage – A review, Journal of Energy Storage 25 (2019) 100852. https://doi.org/10.1016/j.est.2019.100852.

[21]   Md.Y. Bhat, S.A. Hashmi, M. Khan, D. Choi, A. Qurashi, Frontiers and recent developments on supercapacitor's materials, design, and applications: Transport and power system applications, Journal of Energy Storage 58 (2023) 106104. https://doi.org/10.1016/j.est.2022.106104.

[22]   F. Ahmad, A. Shahzad, M. Danish, M. Fatima, M. Adnan, S. Atiq, M. Asim, M.A. Khan, Q.U. Ain, R. Perveen, Recent developments in transition metal oxide-based electrode composites for supercapacitor applications, Journal of Energy Storage 81 (2024) 110430. https://doi.org/10.1016/j.est.2024.110430.

[23]   H. Wang, L. Sheng, G. Yasin, L. Wang, H. Xu, X. He, Reviewing the current status and development of polymer electrolytes for solid-state lithium batteries, Energy Storage Materials 33 (2020) 188–215. https://doi.org/10.1016/j.ensm.2020.08.014.

[24]   C. Zhong, Y. Deng, W. Hu, J. Qiao, L. Zhang, J. Zhang, A review of electrolyte materials and compositions for electrochemical supercapacitors, Chem. Soc. Rev. 44 (2015) 7484–7539. https://doi.org/10.1039/C5CS00303B.

[25]   C. Zhao, W. Zheng, A Review for Aqueous Electrochemical Supercapacitors, Front. Energy Res. 3 (2015). https://doi.org/10.3389/fenrg.2015.00023.

[26]   Md.Y. Bhat, N. Yadav, S.A. Hashmi, Gel Polymer Electrolyte Composition Incorporating Adiponitrile as a Solvent for High-Performance Electrical Double-Layer Capacitor, ACS Appl. Energy Mater. 3 (2020) 10642–10652. https://doi.org/10.1021/acsaem.0c01690.

[27]   S.A. Hashmi, N. Yadav, M.K. Singh, Polymer Electrolytes for Supercapacitor and Challenges, in: Polymer Electrolytes, John Wiley & Sons, Ltd, 2020: pp. 231–297. https://doi.org/10.1002/9783527805457.ch9.

[28]   D. Wei, S.J. Wakeham, T.W. Ng, M.J. Thwaites, H. Brown, P. Beecher, Transparent, flexible and solid-state supercapacitors based on room temperature ionic liquid gel, Electrochemistry Communications 11 (2009) 2285–2287. https://doi.org/10.1016/j.elecom.2009.10.011.

[29]   M. Naguib, M. Kurtoglu, V. Presser, J. Lu, J. Niu, M. Heon, L. Hultman, Y. Gogotsi, M.W. Barsoum, Two-Dimensional Nanocrystals Produced by Exfoliation of Ti3AlC2, Advanced Materials 23 (2011) 4248–4253. https://doi.org/10.1002/adma.201102306.

[30]   L. Verger, V. Natu, M. Carey, M.W. Barsoum, MXenes: An Introduction of Their Synthesis, Select Properties, and Applications, Trends in Chemistry 1 (2019) 656–669. https://doi.org/10.1016/j.trechm.2019.04.006.

[31]   Two-Dimensional Nanocrystals Produced by Exfoliation of Ti3AlC2 - Naguib - 2011 - Advanced Materials - Wiley Online Library, (n.d.). https://onlinelibrary.wiley.com/doi/10.1002/adma.201102306 (accessed April 23, 2024).

[32]   S. Abdolhosseinzadeh, X. Jiang, H. Zhang, J. Qiu, C. (John) Zhang, Perspectives on solution processing of two-dimensional MXenes, Materials Today 48 (2021) 214–240. https://doi.org/10.1016/j.mattod.2021.02.010.

[33]   Ti3C2 MXene: recent progress in its fundamentals, synthesis, and applications | Rare Metals, (n.d.). https://link.springer.com/article/10.1007/s12598-022-02058-2 (accessed April 24, 2024).

[34]   L. Ma, L.R.L. Ting, V. Molinari, C. Giordano, B.S. Yeo, Efficient hydrogen evolution reaction catalyzed by molybdenum carbide and molybdenum nitride nanocatalysts synthesized via the urea glass route, J. Mater. Chem. A 3 (2015) 8361–8368. https://doi.org/10.1039/C5TA00139K.

[35]   P. Urbankowski, B. Anasori, T. Makaryan, D. Er, S. Kota, P.L. Walsh, M. Zhao, V.B. Shenoy, M.W. Barsoum, Y. Gogotsi, Synthesis of two-dimensional titanium nitride Ti4N3 (MXene), Nanoscale 8 (2016) 11385–11391. https://doi.org/10.1039/C6NR02253G.

[36]   C. Xu, L. Wang, Z. Liu, L. Chen, J. Guo, N. Kang, X.-L. Ma, H.-M. Cheng, W. Ren, Large-area high-quality 2D ultrathin Mo2C superconducting crystals, Nature Mater 14 (2015) 1135–1141. https://doi.org/10.1038/nmat4374.

[37]   T. Li, L. Yao, Q. Liu, J. Gu, R. Luo, J. Li, X. Yan, W. Wang, P. Liu, B. Chen, W. Zhang, W. Abbas, R. Naz, D. Zhang, Fluorine-Free Synthesis of High-Purity Ti3C2Tx (T=OH, O) via Alkali Treatment, Angewandte Chemie International Edition 57 (2018) 6115–6119. https://doi.org/10.1002/anie.201800887.

[38]   S. Yang, P. Zhang, F. Wang, A.G. Ricciardulli, M.R. Lohe, P.W.M. Blom, X. Feng, Fluoride-Free Synthesis of Two-Dimensional Titanium Carbide (MXene) Using A Binary Aqueous System, Angewandte Chemie 130 (2018) 15717–15721. https://doi.org/10.1002/ange.201809662.

[39]   W. Sun, S.A. Shah, Y. Chen, Z. Tan, H. Gao, T. Habib, M. Radovic, M.J. Green, Electrochemical etching of Ti2AlC to Ti2CTx (MXene) in low-concentration hydrochloric acid solution, J. Mater. Chem. A 5 (2017) 21663–21668. https://doi.org/10.1039/C7TA05574A.

[40]   X.-H. Zha, K. Luo, Q. Li, Q. Huang, J. He, X. Wen, S. Du, Role of the surface effect on the structural, electronic and mechanical properties of the carbide MXenes, EPL 111 (2015) 26007. https://doi.org/10.1209/0295-5075/111/26007.

[41]   K.A. Papadopoulou, A. Chroneos, D. Parfitt, S.-R.G. Christopoulos, A perspective on MXenes: Their synthesis, properties, and recent applications, Journal of Applied Physics 128 (2020) 170902. https://doi.org/10.1063/5.0021485.

[42]   C.J. Zhang, S. Pinilla, N. McEvoy, C.P. Cullen, B. Anasori, E. Long, S.-H. Park, A. Seral-Ascaso, A. Shmeliov, D. Krishnan, C. Morant, X. Liu, G.S. Duesberg, Y. Gogotsi, V. Nicolosi, Oxidation Stability of Colloidal Two-Dimensional Titanium Carbides (MXenes), Chem. Mater. 29 (2017) 4848–4856. https://doi.org/10.1021/acs.chemmater.7b00745.

[43]   N. Zhang, Y. Hong, S. Yazdanparast, M.A. Zaeem, Superior structural, elastic and electronic properties of 2D titanium nitride MXenes over carbide MXenes: a comprehensive first principles study, 2D Mater. 5 (2018) 045004. https://doi.org/10.1088/2053-1583/aacfb3.

[44]   H. Kumar, N.C. Frey, L. Dong, B. Anasori, Y. Gogotsi, V.B. Shenoy, Tunable Magnetism and Transport Properties in Nitride MXenes, ACS Nano 11 (2017) 7648–7655. https://doi.org/10.1021/acsnano.7b02578.

[45]   A. Bandyopadhyay, D. Ghosh, S.K. Pati, Effects of point defects on the magnetoelectronic structures of MXenes from first principles, Phys. Chem. Chem. Phys. 20 (2018) 4012–4019. https://doi.org/10.1039/C7CP07165E.

[46]   Y. Dall'Agnese, M.R. Lukatskaya, K.M. Cook, P.-L. Taberna, Y. Gogotsi, P. Simon, High capacitance of surface-modified 2D titanium carbide in acidic electrolyte, Electrochemistry Communications 48 (2014) 118–122. https://doi.org/10.1016/j.elecom.2014.09.002.

[47]   U.M. Patil, S.B. Kulkarni, V.S. Jamadade, C.D. Lokhande, Chemically synthesized hydrous RuO2 thin films for supercapacitor application, Journal of Alloys and Compounds 509 (2011) 1677–1682. https://doi.org/10.1016/j.jallcom.2010.09.133.

[48]   P.J. Sephra, P. Baraneedharan, C. Tharini, 5 - Transition metal oxides/sulfides electrode–based supercapacitors, in: S.G. Krishnan, H.D. Pham, D.P. Dubal (Eds.), Supercapacitors, Elsevier, 2024: pp. 93–123. https://doi.org/10.1016/B978-0-443-15478-2.00009-7.

[49]   F. Li, Y.-L. Liu, G.-G. Wang, H.-Y. Zhang, B. Zhang, G.-Z. Li, Z.-P. Wu, L.-Y. Dang, J.-C. Han, Few-layered Ti3C2Tx MXenes coupled with Fe2O3 nanorod arrays grown on carbon cloth as anodes for flexible asymmetric supercapacitors, J. Mater. Chem. A 7 (2019) 22631–22641. https://doi.org/10.1039/C9TA08144E.

[50]   Y. Tian, C. Yang, W. Que, X. Liu, X. Yin, L.B. Kong, Flexible and free-standing 2D titanium carbide film decorated with manganese oxide nanoparticles as a high volumetric capacity electrode for supercapacitor, Journal of Power Sources 359 (2017) 332–339. https://doi.org/10.1016/j.jpowsour.2017.05.081.

[51]   R.A. Chavan, G.P. Kamble, S.B. Dhavale, A.S. Rasal, S.S. Kolekar, J.-Y. Chang, A.V. Ghule, NiO@MXene Nanocomposite as an Anode with Enhanced Energy Density for Asymmetric Supercapacitors, Energy Fuels 37 (2023) 4658–4670. https://doi.org/10.1021/acs.energyfuels.2c04206.

[52]   J. Vigneshwaran, R.L. Narayan, D. Ghosh, V. Chakkravarthy, S.P. Jose, Robust hierarchical three dimensional nickel cobalt tungstate-MXene nanocomposite for high performance symmetric coin cell supercapacitors, Journal of Energy Storage 56 (2022) 106102. https://doi.org/10.1016/j.est.2022.106102.

[53]   Q. Xia, W. Cao, F. Xu, Y. Liu, W. Zhao, N. Chen, G. Du, Assembling MnCo2O4 nanoparticles embedded into MXene with effectively improved electrochemical performance, Journal of Energy Storage 47 (2022) 103906. https://doi.org/10.1016/j.est.2021.103906.

[54]   W. Zheng, J. Halim, A.S. Etman, A.E. Ghazaly, J. Rosen, M.W. Barsoum, Boosting the volumetric capacitance of MoO3-x free-standing films with Ti3C2 MXene, Electrochimica Acta 370 (2021) 137665. https://doi.org/10.1016/j.electacta.2020.137665.

[55]   W. Zheng, J. Halim, A. El Ghazaly, A.S. Etman, E.N. Tseng, P.O.Å. Persson, J. Rosen, M.W. Barsoum, Flexible Free-Standing MoO3/Ti3C2Tz MXene Composite Films with High Gravimetric and Volumetric Capacities, Advanced Science 8 (2021) 2003656. https://doi.org/10.1002/advs.202003656.

[56]   Q. Wang, Z. Zhang, Z. Zhang, X. Zhou, G. Ma, Facile synthesis of MXene/MnO2 composite with high specific capacitance, J Solid State Electrochem 23 (2019) 361–365. https://doi.org/10.1007/s10008-018-4143-4.

[57]   R.B. Rakhi, B. Ahmed, D. Anjum, H.N. Alshareef, Direct Chemical Synthesis of MnO2 Nanowhiskers on Transition-Metal Carbide Surfaces for Supercapacitor Applications, ACS Appl. Mater. Interfaces 8 (2016) 18806–18814. https://doi.org/10.1021/acsami.6b04481.

[58]   H. Jiang, Z. Wang, Q. Yang, M. Hanif, Z. Wang, L. Dong, M. Dong, A novel MnO2/Ti3C2Tx MXene nanocomposite as high performance electrode materials for flexible supercapacitors, Electrochimica Acta 290 (2018) 695–703. https://doi.org/10.1016/j.electacta.2018.08.096.

[59]   S. Chen, Y. Xiang, W. Xu, C. Peng, A novel MnO2/MXene composite prepared by electrostatic self-assembly and its use as an electrode for enhanced supercapacitive performance, Inorg. Chem. Front. 6 (2019) 199–208. https://doi.org/10.1039/C8QI00957K.

[60]   W. Yuan, L. Cheng, B. Zhang, H. Wu, 2D-Ti3C2 as hard, conductive substrates to enhance the electrochemical performance of MnO2 for supercapacitor applications, Ceramics International 44 (2018) 17539–17543. https://doi.org/10.1016/j.ceramint.2018.06.086.

[61]   R. Ramachandran, C. Zhao, M. Rajkumar, K. Rajavel, P. Zhu, W. Xuan, Z.-X. Xu, F. Wang, Porous nickel oxide microsphere and Ti3C2Tx hybrid derived from metal-organic framework for battery-type supercapacitor electrode and non-enzymatic H2O2 sensor, Electrochimica Acta 322 (2019) 134771. https://doi.org/10.1016/j.electacta.2019.134771.

[62]   Q.X. Xia, J. Fu, J.M. Yun, R.S. Mane, K.H. Kim, High volumetric energy density annealed-MXene-nickel oxide/MXene asymmetric supercapacitor, RSC Adv. 7 (2017) 11000–11011. https://doi.org/10.1039/C6RA27880A.

[63]   N.A. Althubiti, S. Aman, T.A.M. Taha, Synthesis of MnFe2O4/MXene/NF nanosized composite for supercapacitor application, Ceramics International 49 (2023) 27496–27505. https://doi.org/10.1016/j.ceramint.2023.06.025.

[64]   J. Zhu, Y. Tang, C. Yang, F. Wang, M. Cao, Composites of TiO2 Nanoparticles Deposited on Ti3C2 MXene Nanosheets with Enhanced Electrochemical Performance, J. Electrochem. Soc. 163 (2016) A785. https://doi.org/10.1149/2.0981605jes.

[65]   T. Arun, A. Mohanty, A. Rosenkranz, B. Wang, J. Yu, M.J. Morel, R. Udayabhaskar, S.A. Hevia, A. Akbari-Fakhrabadi, R.V. Mangalaraja, A. Ramadoss, Role of electrolytes on the electrochemical characteristics of Fe3O4/MXene/RGO composites for supercapacitor applications, Electrochimica Acta 367 (2021) 137473. https://doi.org/10.1016/j.electacta.2020.137473.

[66]   S.B. Ambade, R.B. Ambade, W. Eom, S.H. Noh, S.H. Kim, T.H. Han, 2D Ti3C2 MXene/WO3 Hybrid Architectures for High-Rate Supercapacitors, Advanced Materials Interfaces 5 (2018) 1801361. https://doi.org/10.1002/admi.201801361.

[67]   I. Ayman, A. Rasheed, S. Ajmal, A. Rehman, A. Ali, I. Shakir, M.F. Warsi, CoFe2O4 Nanoparticle-Decorated 2D MXene: A Novel Hybrid Material for Supercapacitor Applications,Energy Fuels 34 (2020) 7622–7630. https://doi.org/10.1021/acs.energyfuels.0c00959.

[68]   F. Wang, M. Cao, Y. Qin, J. Zhu, L. Wang, Y. Tang, ZnO nanoparticle-decorated two-dimensional titanium carbide with enhanced supercapacitive performance, RSC Adv. 6 (2016) 88934–88942. https://doi.org/10.1039/C6RA15384D.

[69]   M. Mahmood, K. Chaudhary, M. Shahid, I. Shakir, P.O. Agboola, M. Aadil, Fabrication of MoO3 Nanowires/MXene@CC hybrid as highly conductive and flexible electrode for

next-generation supercapacitors applications, Ceramics International 48 (2022)19314–19323. https://doi.org/10.1016/j.ceramint.2022.03.226.

[70]   J. Zhu, X. Lu, L. Wang, Synthesis of a MoO3/Ti3C2Tx composite with enhanced capacitive performance for supercapacitors, RSC Adv. 6 (2016) 98506–98513. https://doi.org/10.1039/C6RA15651G.

[71]   X. Zhang, B. Shao, A. Guo, Z. Gao, Y. Qin, C. Zhang, F. Cui, X. Yang, Improved electrochemical performance of CoOx-NiO/Ti3C2Tx MXene nanocomposites by atomic layer deposition towards high capacitance supercapacitors, Journal of Alloys and Compounds 862 (2021) 158546. https://doi.org/10.1016/j.jallcom.2020.158546.

[72]   C. Li, G. Jiang, T. Liu, Z. Zeng, P. Li, R. Wang, X. Zhang, NiCoO2 nanosheets interlayer network connected in reduced graphene oxide and MXene for high-performance asymmetric supercapacitors, Journal of Energy Storage 49 (2022) 104176. https://doi.org/10.1016/j.est.2022.104176.

[73]   Y. Zhang, J. Cao, Z. Yuan, L. Zhao, L. Wang, W. Han, Assembling Co3O4 Nanoparticles into MXene with Enhanced electrochemical performance for advanced asymmetric supercapacitors, Journal of Colloid and Interface Science 599 (2021) 109–118. https://doi.org/10.1016/j.jcis.2021.04.089.

[74]   C. Zhang, M. Beidaghi, M. Naguib, M.R. Lukatskaya, M.-Q. Zhao, B. Dyatkin, K.M. Cook, S.J. Kim, B. Eng, X. Xiao, D. Long, W. Qiao, B. Dunn, Y. Gogotsi, Synthesis and Charge Storage Properties of Hierarchical Niobium Pentoxide/Carbon/Niobium Carbide (MXene) Hybrid Materials, Chem. Mater. 28 (2016) 3937–3943. https://doi.org/10.1021/acs.chemmater.6b01244.

[75]   X. Lu, J. Zhu, W. Wu, B. Zhang, Hierarchical architecture of PANI@TiO2/Ti3C2Tx ternary composite electrode for enhanced electrochemical performance, Electrochimica Acta 228 (2017) 282–289. https://doi.org/10.1016/j.electacta.2017.01.025.

[76]   C. Peng, Z. Kuai, T. Zeng, Y. Yu, Z. Li, J. Zuo, S. Chen, S. Pan, L. Li, WO3 Nanorods/MXene composite as high performance electrode for supercapacitors, Journal of Alloys and Compounds 810 (2019) 151928. https://doi.org/10.1016/j.jallcom.2019.151928.

[77]   J. Zhao, F. Liu, W. Li, Phosphate Ion-Modified RuO2/Ti3C2 Composite as a High-Performance Supercapacitor Material, Nanomaterials (Basel) 9 (2019) 377. https://doi.org/10.3390/nano9030377.

[78]   Y. Wang, J. Sun, X. Qian, Y. Zhang, L. Yu, R. Niu, H. Zhao, J. Zhu, 2D/2D heterostructures of nickel molybdate and MXene with strong coupled synergistic effect towards enhanced supercapacitor performance, Journal of Power Sources 414 (2019) 540–546. https://doi.org/10.1016/j.jpowsour.2019.01.036.

[79]   J. Song, P. Hu, Y. Liu, W. Song, X. Wu, Enhanced Electrochemical Performance of Co2NiO4/Ti3C2Tx Structures through Coupled Synergistic Effects, ChemistrySelect 4 (2019) 12886–12890. https://doi.org/10.1002/slct.201903511.

Emerging Materials for Next Frontier Energy and Environment Applications   Materials Research Forum LLC
Materials Research Foundations 170 (2024) 145-167   https://doi.org/10.21741/9781644903292-7

Chapter 7

Synergistic effects in MXene: Transition metal chalcogenides to unlock supercapacitor potential

Harshini Sharan, Pavithra Karthikesan, Jayachandran Madhavan, Alagiri Mani[*]

Department of Physics and Nanotechnology, SRM Institute of Science and Technology, Kattankulathur, Chengalpattu, Tamil Nadu - 603 203, India

[*] alagirim@srmist.edu.in

Abstract

The foremost nations and scientific circles are focusing the spotlight mainly on energy because of the shifting global environment. The development and upgrade of energy storage systems featuring higher efficiency have drawn a great deal of concern. Therefore, significant advancements in energy storage might be made possible by a high-power density device called a supercapacitor (SCs). The use of 2D layered materials has sparked a lot of attention in the modern era since these materials have suitable electrochemical and physiochemical properties that make them ideal for high-performance energy storage devices. Amidst several 2D materials, MXenes have garnered significant interest due to their hydrophilic nature, metallic conductivity, rich active sites, and high surface area. In virtue of these upsides, compositing MXene with electrochemically favourable Ternary Metal Chalcogenides (TMCs) electrode material will significantly improve the composite performance by reducing the agglomeration of nanoparticles and inhibit restacking of MXene sheets through synergistic effects. This chapter mainly focuses on the MXene-TMCs composites for supercapacitor application and discusses the process involving them.

Keywords

MXene, Ternary Metal Chalcogenides, Nanocomposites, Supercapacitors, 2D Materials

Contents

Synergistic effects in MXene: Transition metal chalcogenides to unlock supercapacitor potential 145
1.   Introduction...146
2.   The check list of energy storage mechanism 147
    2.1   Capacitive and non-capacitive methods of energy storage 147
    2.2   Contrast between three types of supercapacitors...149
3.   Transition metal chalcogenides (TMCs) as an efficient electrode material for supercapacitors 151

3.1   Transition metal sulphides (TMSs) 152

3.2   Transition metal selenides (TMSe's) 152

3.3   Transition metal tellurides (TMTs) 153

4.   TMCs/MXene: Recent progress **154**

4.1   MXene-An emerging trait 154

4.2   Transition metal sulphides/ MXene (TMSs/MXene) 156

4.3   Transition metal selenides/ MXene (TMSe's/MXene) 157

4.4   Transition metal tellurides/MXene (TMT's/MXene) 158

Conclusion **159**

References **160**

1.   Introduction

Energy is an anomaly on the agenda in numerous aspects of the 21st century, assessing the world's economy, one's self, and the mass order. It is noteworthy mainly because of its key attributes, including an infinite resource, replacement, supply prices, and everlasting attributes. Making the switch towards environment friendly and endless sources of energy has grown to be a key idea for societies globally because it facilitates productive generation and use of traditional energy sources despite cutting down on the adverse environmental impacts of fossil fuel-based energies. Such an evolution is vital to building an ecologically sound global sequence and calls for the launch of renewable energy sources and alternative energy strategies. Subsequently, over the period of the industrial era, most notably post-World War II, Western economies witnessed tremendous growth in their economies and an upsurge in consumption of energy. This upward trend continued to grow with the outbreak of the period known as the Cold war. The myth of "Global warming", which at first acquired prominence in the 2000s, is cautiously acknowledged by numerous nations, both industrialized and developing. However, it is evident that the uncontrolled release of vast amounts of carbon during energy production, causing global warming, will soon lead to significant global challenges. The threats of global warming and contamination of the environment originate with the heavy reliance on fossil fuels to satisfy an important percentage of the global energy demand. Furthermore, concerns about the potential depletion of fossil fuel reserves soon led to a great deal of study on renewable sources, which could be generated via the wind, hydrogen, sun, and water, originates from the current energy flow in periodically recurring innate actions. Nevertheless, uncertain about the inevitable destruction of supplies like oil, coal, and other fuels derived from petroleum and natural gas, there are still suspicious concerning a finite lifespan. In turn, these sources of energy are categorized as non-renewable and unsustainable, being that more than 50% of the global need for energy is currently met by such sources. One is aware of an enthusiasm for renewable energy sources. In the year 1975, the portion of power generated utilizing low carbon emission technologies for global electricity production was 21.5%. As of 2020, this figure had risen to 39%. Unquestionably, this rate surged as a result of the measures implemented to address the climate catastrophe, particularly the Kyoto Agreement. Meanwhile, both consumption and supply of energy spotted dramatic increases and had been extremely divisive. Furthermore, the proportion of solar energy, which had been negligible in the 1990s, surged to 4% in 2020, while the portion of wind energy rose to 6.7%. We should not handle the energy problem casually, given the depletion of fossil fuels and the unprecedented emission of greenhouse gases into the environment [1]. In light of this issue, energy storage devices, especially SCs, have sparked substantial interest from researchers and organizations. In this scenario, a supercapacitor is a cutting-edge energy storage technology that can store significant capacitance in a compact space. It finds vast application in an array of electrical gadgets. This sort of technology holds multiple benefits, including its capacity to function well in a broad range of temperatures, an eternal life span, an effortless rechargeable circuit, as well as rapid charging capacities, and affordability [2]. Furthermore, SCs have been manufactured and utilized for high-power applications for decades, as opposed to rechargeable batteries and will be included into all energy conversion devices in the days to come this in turn, provides energy even when resources are few. The worldwide SC market has boosted the use of regenerative braking systems in elevators and hybrid electric vehicles (HEVs), as well as the demand for SCs in energy harvesting applications and the use of SCs in locomotive applications such as trains and airplanes has gone high. The market for electrolytes subject to guidelines and regulations has been analyzed using electrolyte selection from SC manufacturers. The annual global market for SCs and its future predictions has described in Fig. 1 [3]. This chapter deals with the exploration of transition metal chalcogenides, namely sulphides, selenides and, tellurides with their MXene nanocomposites for enhanced supercapacitor performance.

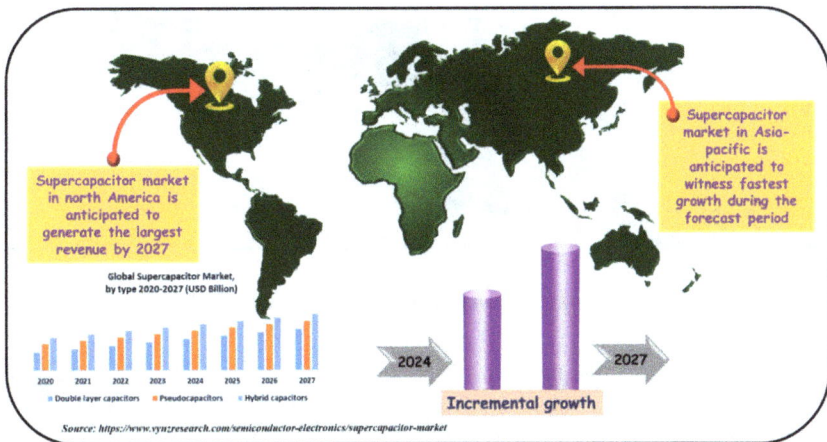

Fig. 1 Global SC market by types till 2027.

2. The check list of energy storage mechanism

2.1 Capacitive and non-capacitive methods of energy storage

In the realm of commercial electronics, supercapacitors (SCs), also known as ultracapacitors, are widely sought after due to their high-power densities and life span, as well as their simplistic design and operation mechanism. Fig. 2 [4] shows the progress, evolution and the development of SCs. Electrostatic interactions or chemical processes limited to the interface enable these devices to store charge. One of the characteristics of a SC is its potential, which acts as a measure of the amount of charge that it possesses and indicates a continuous shift in free energy in tandem with physical transformation [5].

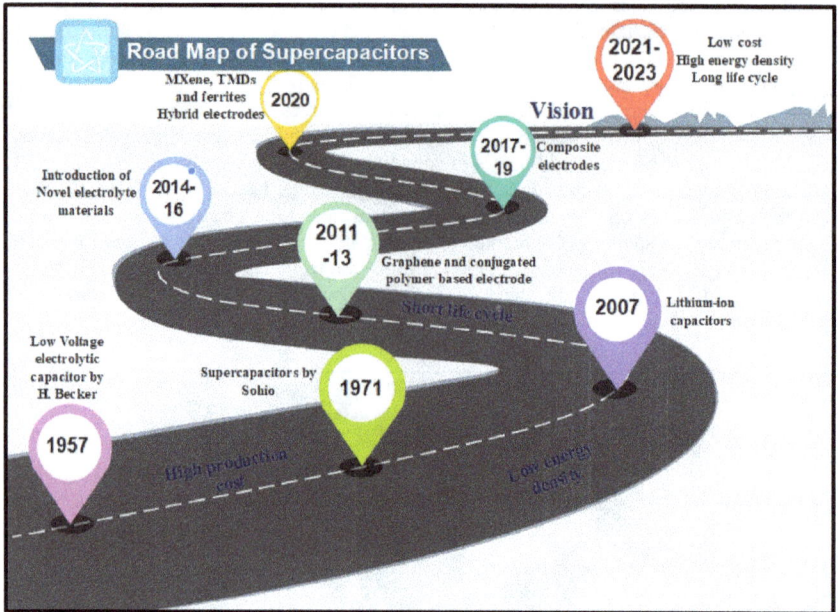

Fig. 2 The process, evolution and the development of SCs [4]

Fig. 3 illustrates the categorization of charge storage mechanisms. Based on charge storage by electrode materials, SCs are classed as electrical-double layered capacitors (EDLC), pseudo-capacitors, and hybrid SCs. Amongst, EDLCs storing charge electrostatically, almost pseudocapacitors and EDLCs are two distinct kinds of capacitors that differ based on charge storage. The Pseudocapacitors store energy through either a pseudo-intercalation type reaction (EDLCs) or a rapid faradaic redox reaction occurring at the electrode surface, where charge is formed in the two layers. Conversely, battery-type materials experience performance-enhancing reactions that are exclusively faradaic in nature. Hybrid supercapacitors store charge in two ways: electrostatically and electrochemically, combining the advantages of both EDLCs and pseudocapacitors [6].

Emerging Materials for Next Frontier Energy and Environment Applications   Materials Research Forum LLC
Materials Research Foundations 170 (2024) 145-167   https://doi.org/10.21741/9781644903292-7

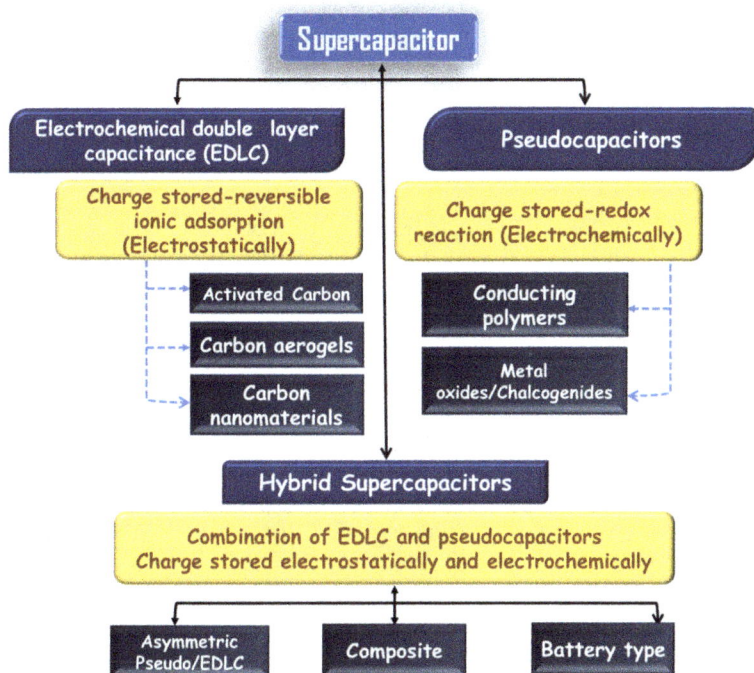

Fig. 3 Classification of SC's and their electrode materials

2.2   Contrast between three types of supercapacitors

Having an in-depth overview of the operating procedures of energy-storage devices is essential, not merely from a basic perspective but also in terms of ion storage and transportation mechanisms. This knowledge serves as a solid foundation for the development of gadgets and their potential functionality.

Electrochemical Double Layer Capacitors (EDLCs) possess two methods of charge storage: non-faradic storage, which doesn't entail any charge transfer between the electrode and the electrolyte, or electrostatic storage. Under the influence of an applied voltage, the ions in the electrolyte migrate across the separator and adhere to the pores of the electrode that carry an opposing charge. This phenomenon occurs due to the presence of a potential difference, which exerts a force that attracts charges of opposite polarity towards one another. A bilayer of charge forms at the electrodes to avoid ion recombination. EDLCs may achieve greater energy densities due to the existence of a double layer, increased surface area, and decreased distances between electrodes. EDLCs are fabricated by combining two carbon-based components: an electrolyte and a separator [7].

Pseudo capacitors (PCs) store the electric charge by means of faradic mechanisms, which include the charge flow between the electrode and electrolyte interface. Suddenly, when an electric potential is supplied to a pseudocapacitors, the electrode material undergoes reduction and oxidation processes. These reactions comprise the transfer of electric charge across the double layer which results in a stream of faradic current over the supercapacitor cell. This type of SC uses the faradic mechanism to attain higher specific capacitance and energy densities in comparison to EDLCs. Examples are metal oxides, metal chalcogenides (Sulphide, selenides and tellurides), metal nitrides, conducting polymers, etc.

Electric double-layer capacitors (EDLCs) provide exceptional durability and power efficiency, whereas pseudo capacitance exhibits an even greater capacity for storing electrical charge. A hybrid system combines a battery-like electrode and a capacitor-like electrode, one for storing energy and one for generating power inside a single cell. One electrode adopts the double-layer storage mechanism, namely porous carbon, while the second electrode stores charge using faradaic processes; such processes were seen in transition metal oxides and, transition metal chalcogenides, etc. By using a suitable combination of electrodes, it is possible to amplify the voltage of the cell, leading to enhanced power and energy densities [8]. The charge storage mechanism of three types of SCs are displayed in Fig. 4.

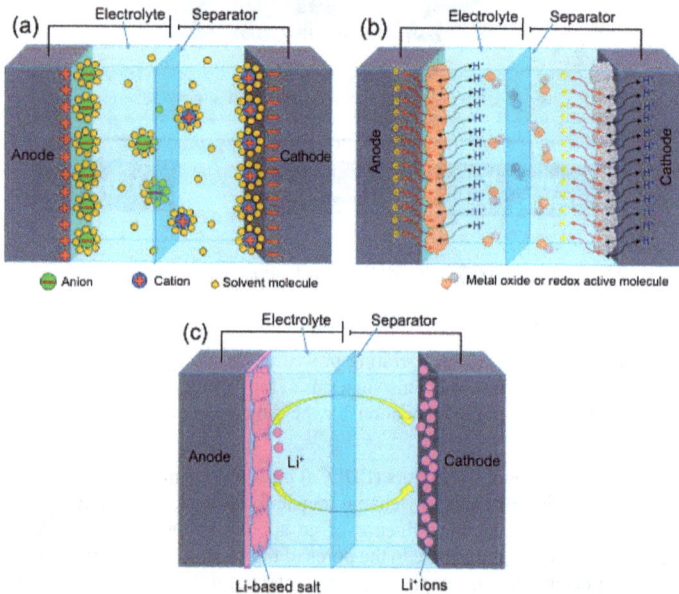

Fig. 4 Schematic representation of three types of SCs (a) EDLC, (b) Pseudocapacitors and, (c) Hybrid capacitor, reproduced from [9] an open access article

Emerging Materials for Next Frontier Energy and Environment Applications   Materials Research Forum LLC
Materials Research Foundations 170 (2024) 145-167   https://doi.org/10.21741/9781644903292-7

3.   Transition metal chalcogenides (TMCs) as an efficient electrode material for supercapacitors

TMCs are a prominent group of crystalline materials composed of two layers that mix transition metal elements from groups IV to VIIB of the periodic chart with group VIA metals such as sulphur, selenium, and, tellurium which have diverse magnetic, catalytic, optical, and electrical characteristics. TMCs have the standardized formula AX_2, wherein A is a transition metal and X is a chalcogen. TMCs tend to possess bandgaps ranging from 0 to 2 eV and, their crystalline structures are similar to graphene's zero bandgap. The formation of layered structures in TMCs occurs due to the arrangement of hexagonally packed planes. In these layered structures, the coordination surrounding the metal groups can either be trigonal, prismatic, or octahedral. The electrochemical characteristics of TMCs, including their abundant intercalation sites, extensive electrochemically active surface areas, few diffusion routes, excellent electrical conductivities, and enhanced material stability, may be effectively harnessed for applications in SCs [10]. These are materials with layered structures similar to graphite and Van der Waals-type crystal structures, hold significant potential as electrode materials for SCs due to their exceptional ability to undergo a broad variety of oxidation states related to transition metals and possess exceptional Faradaic charge-storing abilities. Edge-oriented TMCs are more fascinating than nanostructures dominated by basal planes because they provide a higher concentration of electrochemically active regions as well as open spaces between layers for efficient diffusion and intercalation of electrolyte ions [11]. In such cases, the fabrication of binary TMCs (AX_2; A=transition metal elements, X=S, Se, Te), ternary TMCs (ABX_2; AB=transition metal elements, X=S, Se, Te), and their composites with carbon-based materials has sparked a great deal of interest among the researcher communities. Therefore, TMCs play a dynamic role in the field of energy storage, specifically SCs. Fig. 5. Shows the schematic of TMCs and the inter/intra layer representation of forces.

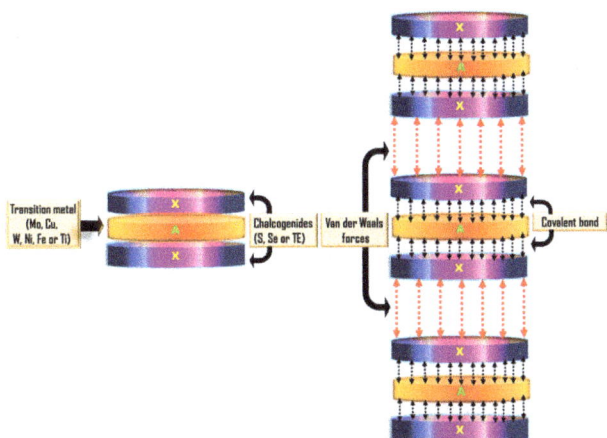

Fig. 5 Schematic of TMCs with the inter/intra layer representation of forces, Reproduced from. [4] with permission from the Royal Society of Chemistry.

Emerging Materials for Next Frontier Energy and Environment Applications   Materials Research Forum LLC
Materials Research Foundations 170 (2024) 145-167   https://doi.org/10.21741/9781644903292-7

3.1   Transition metal sulphides (TMSs)

TMSs nanomaterials have drawn huge interest from experts dealing with material science and nanotechnology. This pertains to their uniquely designed structural, physical, and chemical features, which have made them ideal to be employed as prospective materials in a wide variety of optoelectronic devices such as gadgets, indicators, superconductive materials, solar cells, as well as sensors. In addition to these features, it also exhibits high faradaic electrochemical behaviour as well as high theoretical capacities that can be used to make cost-effective and eco-friendly technologies. Fortunately, metal oxides and carbon-based materials have usually ruled research in this domain; nonetheless, TMSs have recently captured the spotlight owing to their superior electrical conductivity and reliable thermal and mechanical properties. Added to that, they enhance ion transportation by enabling the movement of sulphur atoms, which sets them apart from similar materials. Mostly, these classes of materials are harmless and abundant, making them economically viable for large-scale production [11,12,13]. Using sulphide as electrodes requires meticulous investigation of their size, structure, and composition to provide robust electrochemical performance via reduced diffusion routes and improved conductivity. At this point, a handful of synthesis techniques have been used to tailor the morphologies to achieve the desired activity, including microwave, ball milling, atomic and chemical layer deposition, hydrothermal and solvothermal processes, electrodeposition, and finally electrochemical synthesis [14,15]. This prospect paved the way for designing binary and ternary transition metal sulphide electrode materials for SCs. On this occasion, MoS_2 and WS_2 are possibly the most deliberate layered TMS materials with enhanced supercapacitor performance. Wei et al. [16] described self-assembled MoS_2 nanoflowers via facile hydrothermal method and achieved a greater specific capacitance of about 1120 F/g with a retention of 96% after 2000 cycles. Likewise, Nagaraju et al. [17] reported the hydrothermal synthesis of WS_2 which shows a specific capacitance of 1439.5 F/g at 5 A/g current density and, 77.4% retention after 3000 cycles. Furthermore, Krishna et al. [18] prepared $CoS@WS_2$ composite electrodes through a simple hydrothermal route and obtained an exemplary specific capacitance of 2442.32 F/g at 4.28 A/g current density and exhibits 97.16% retention over 3000 cycles, respectively. For example, in the case of ternary TMSs, Wang et al. [19] via a feasible hydrothermal method synthesized $NiCo_2S_4$ nanorods and exhibits excellent electrochemical behaviour having a 3093 F/g specific capacitance at the current density of 5 A/g. In addition to this, the asymmetric device was fabricated, displaying a 39.3 Wh/kg energy density at a power density of 800 W/kg, respectively.

3.2   Transition metal selenides (TMSe's)

Transition metal selenides mimic transition metal sulphides as selenium falls into the same group as sulphur. However, these classes of materials have gained attention for their distinctive electrical and optical characteristics, which makes them promising materials for use in solar cells, electrocatalysis and photocatalysis. Added to that, the metal-S and metal-O bonds are stronger than the metal-Se bonds, making them preferable for conversion reactions [20] TMSe's may serve as an electrode material owing to its greater volume of energy density and rate capability than its sulphides counterpart. Hence, these materials aroused the appetite of scientists as potential electrode materials for electrochemical energy storage devices. TMSe's often consist of a sandwich structure, where the metal atom is positioned between two selenide layers. Although atoms in all three layers are bound by covalent bonds, the delicate Vander Waals force is the only bond between the sheets. Furthermore, the reversible intercalation of other atoms in the interchain spaces can result in the formation of intercalates, which facilitate the storage of energy in the stacked conductors [21]. TMSe materials are further classified as binary and ternary TMSe, respectively. Some of the TMSe's having good supercapacitor performance were listed below. For example, in the case of binary TMSe, Liu et al. [22] used the microwave synthesis process to create $CoSe_2$ nanosheets for the purpose of examining their supercapacitive performances and attained a precise capacitance of 333 F/g with a yield of 100.97%. Along with the high capacitance retention even after 25,000 cycles. Correspondingly, Yang et al. [23] reported $NiSe_2$ hierarchical spheres via the hydrothermal method and displayed 572.05 C/g capacitance at 0.5 A/g with 75% retention over 1000 cycles. Nonetheless, Liu et al. [24] by CVD and hydrothermal method, demonstrated $MoSe_2$ nanoflakes, which shows 228.25 C/g capacitance at 1 A/g and, 92% capacitance retention after 5000 cycles. Shinde et al. [25] through chemical oxidation, electrodeposition, and selenization reported $NiSe_2@Cu_2Se$ rose petal-like morphology and showed 1923.12 C/g capacitance at 1 A/g current density with 93.6% capacitance retention for 10000 cycles. In the case of ternary TMSe's, Gopi et al. [26] reported $CoFe_2Se_4$ nanorods and $CoNiSe_2$ microsphere composites, which deliver a supreme specific capacity of 183.4 mAh/g at 1 A/g current density and acquires 99.2% retention after 3000 cycles, respectively.

3.3   Transition metal tellurides (TMTs)

Among TMCs, transition metal tellurides received less attention than sulphides and selenides. While proceeding with the list of electronegativity amidst the chalcogen family tellurides possess lower electronegativity than the other group elements likewise, tellurium displays superior conductivity ($10 \times 10^3 S\ m^{-1}$) in comparison with S ($5 \times 10^{-28} S\ m^{-1}$) and Se ($1 \times 10^3 S\ m^{-1}$) . It also has a greater atomic radius than sulphur and oxygen which leads to its the advanced electrochemical properties [27]. In addition to these features, TMTs hold good mechanical stability, excellent electrical conductivity, extended cycle life, low ionization energy, rapid electron transport, a huge surface area, superior redox active frameworks, as well as high specific capacitance [28,29,30,31]. TMTs have been synthesized into a variety of functional materials for use in the environment, energy storage, and conversion, among other domains. Recently, TMTs based binary and ternary electrode materials for SCs have gained interests among the researchers. For example, Manikandan et al. [32] synthesized CoTe nanorods by a simple hydrothermal technique, which showed a high specific capacity of 170 C/g along with outstanding cyclic stability. Additionally, the fabricated device exhibited 40.7 Wh/kg of energy density and 800 W/kg of power density at 1 A/g current density. This device also displayed excellent stability for 10000 cycles at 30 A/g current density with 85% retention. Similarly, Manikandan et al. [33] reported the one-pot hydrothermal synthesis of NiCoTe nanorods using the reducing agents like cetyltrimethylammonium bromide and ascorbic acid. $Ni_{0.7}Co_{0.3}Te$ NRs have a specific capacitance of 433 C/g at a current density of 0.5 A/g and retain 100% capacity even after 5000 cycles. Additionally, they fabricated a hybrid asymmetric supercapacitor device using $Ni_{0.7}Co_{0.3}Te$ NRs as the positive electrode and orange-peel-derived activated carbon as the negative electrode. After 25,000 cycles, it had a 43 Wh/kg high energy density and 905 W/kg power density and also displays long cyclic stability of 91%. Furthermore, Bhol et al. [34] reported that tellurium nanotubes decorated with CoMgTe microtubes which was synthesized through the wet chemical method and, the device acquired 59.2 $\mu Ah/cm^2$ areal capacity at 6 mA/cm^2 of current density along with an energy density of 42.2 Wh/kg at 6857.1 W/kg of power density with a long-life span of over 15000 cycles. Molaei et al. [35] fabricated flower-like $NiTe_2@CoTe_2$ structures on nickel foam by hydrothermal method and disclosed an excellent capacity of 1388.9 C/g and delivered a desired energy and power densities of about 58.85 Wh/kg and 806.85 W/kg with a capacitance retention of 83.43% at 30 A/g, respectively.

4.   TMCs/MXene: Recent progress

To further attain the desired superior electrochemical performance, scientists have put forward their investigations on hybridizing TMCs with carbonaceous and 2D materials namely, graphene, reduced graphene oxide (rGO), carbon nanotubes (CNTs), single and multi-walled CNTs (SWCNTs and MWCNTs), graphitic carbon nitride (g-C_3N_4) and, MXenes, etc. However, real-life applications of TMC electrode materials confront obstacles such as volumetric fluctuations, breakdown of structure, and nanoparticle aggregation during electrochemical reactions. To avoid these constraints, one fascinating tactic it to include carbon-based and the MXene elements into TMCs structures through combining. This approach has been beneficial for boosting overall performance [36]. Hence, incorporating carbon or MXene in combination with TMCs provides two benefits: (i) hybridization of a carbon coating or support to the chalcogenide surface improves the electrical conductivity and consequently enables rapid electron transfer and its kinetics, (ii) volume change in the active material can result in mechanical stress leading to cracking. Though customized materials could serve as anchors for composite materials by minimizing the same among TMCs during charge and discharge processes. Additionally, they prevent particle agglomeration throughout the electrode. All of these reinforce the electrodes overall strength and cycle durability, which also improves contact between the active material and the electrolyte, which leads to improved electrochemical performance [37]. Further, CNTs have great electrical conductivity and rigidity, which renders them suitable for composite structures. Subsequently, rGO is often used to address inherent problems in TMC composites via its encapsulation role. Its tremendous surface area, strong electrical conductivity, and structural versatility make it a promising candidate for compositing with TMCs [38]. Fascinatingly, MXene is an innovative 2D material distinguished from graphene by its metallic conductivity and hydrophilic traits has unveiled remarkable promises in the realm of SCs. The interaction of MXene with TMCs promotes electrical conductivity, boosts the electrochemically active sites as well as it upgrades the cyclic stability [39]. The results facilitate the development of high-performance electrode materials for supercapacitor devices. Herein, we discuss in depth about the hybridization of TMCs with Mxenes.

4.1   MXene-An emerging trait

An exciting new class of metallic or semiconductor featuring good electrical conductivity is the MXenes, which are two-dimensional (2D) transition metal carbides or nitrides. Most often, they are expressed as $M_{n+1}X_nT_x$, where n is either 1 or 3 and M denotes a transition metal such as Ti, Sc, Cr, V, Zr, Nb, Hf, Mo and Ta), whereas X represents nitrogen or carbon, and T denotes the surface terminal group namely, OH^-, F^- and O_2^- [40].

Emerging Materials for Next Frontier Energy and Environment Applications   Materials Research Forum LLC
Materials Research Foundations 170 (2024) 145-167   https://doi.org/10.21741/9781644903292-7

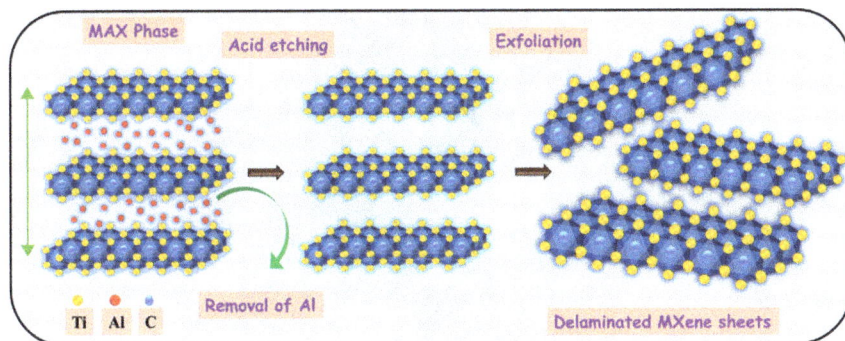

Fig. 6 Synthesis procedure of MXene by acid etching

The synthesis of MXene entails the careful removal of the A-group (i.e., the elements of IIIA and IVA groups) from stacked hexagonal ternary carbides, which are referred to as MAX phases, as shown in Fig. 6. At ambient temperature or higher temperatures, MAX residues are submerged in a regulated amount of hydrofluoric acid (HF) solution. This process results in the emergence of surface depots on MXenes probably they are composed of oxygen and/or hydroxyl groups [41,42]. Owing to the harmful nature of HF, its direct usage is a bit risky. However, in situ generation of HF could be achieved by reacting hydrochloric acid (HCL) with fluoride salts or with the help of ammonium bifluoride. In order to facilitate the parallel intercalation of cations such as Na^+, Li^+, K^+, Ca^+, Al^{3+}, and NH_4^+ across the interlayers, these HF precursors are of great assistance. An added advantage of in situ HF etching over the pure HF etching approach is that it results in delamination without the need for any extra processes. This is because the intercalation of cation and water, which occurs during this process hence, results in delamination. The Etching requirements for the entire transition of the MAX phase to MXene depend on a number of distinct factors. The etching environment is determined by the strength of the M-Al bond in Al-based MAX phases and the rising number n in the $M_{n+1}C_nT_x$ type necessitates etching conditions that are more stringent or longer [43,44]. Despite ongoing investigation on etching techniques, the range of MAX phases that may be carved continues to expand, leading to the advancement of 2D MXene materials. Among the variety of synthesis routes for MXene, Lukatskaya et.al were the pioneers in using MXene as a promising electrode material for supercapacitor application. They discovered that $Ti_3C_2T_x$ MXene generated via HF etching which showed an excellent electrochemical performance [42]. Some of the properties of MXene are described in Fig. 7. In recent times, there has been a boom in utilizing MXene-based electrode materials for SCs because of their aforementioned features. This chapter deals with TMCs/MXene composites for high performance supercapacitor applications.

Emerging Materials for Next Frontier Energy and Environment Applications   Materials Research Forum LLC
Materials Research Foundations 170 (2024) 145-167   https://doi.org/10.21741/9781644903292-7

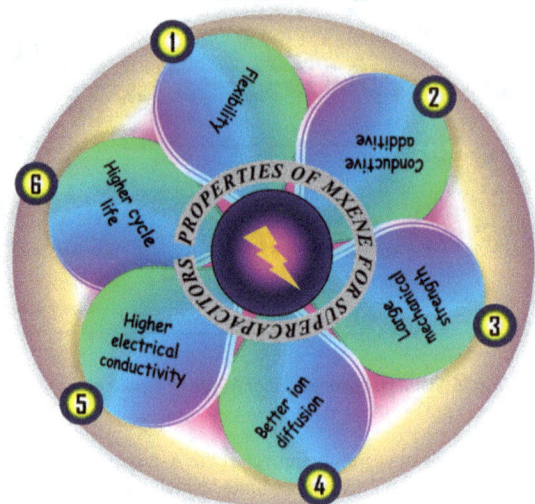

Fig. 7 Properties of MXene for the application of Supercapacitors

4.2   Transition metal sulphides/ MXene (TMSs/MXene)

TMSs are a promising electrode material for SCs due to their higher specific capacitance and improved electrical conductivity compared to metal oxides. Yet, the high-volume shift of TMSs during lithiation often drastically shortens the electrode cycle life. Furthermore, we use MXene as a substrate material to improve conductivity and reduce volume fluctuations during redox processes, thereby providing a buffer area and conductive framework for TMSs. Various TMSs such as CoS_2/MXene, MoS_2/MXene, NiS/MXene, ZnS/MXene, $NiCo_2S_4$/MXene, etc., provide high-performance composite electrode materials for SCs [45,46,47,48,49,50,51,52]. Liu et al. [53] in their study, used ultrasonic treatment to disperse Ti_3C_2 MXene solution. Subsequently, they included cysteine as sulphur source and cobalt chloride as a cobalt precursor in the distributed solution. The solution was then stirred before being transferred into an autoclave and it was maintained at $160°C$ for 6 hrs. Following, solvothermal treatment, the growth of CoS_2 occurs in situ scheduled on the external layer of MXene nanosheets; hence, compounds consisting of CoS_2/MXene are prepared. With this particular tandem, MXene can effectually hinder the volume expansion of CoS_2 and thus showed enhanced conductivity. Meanwhile the CoS_2 that is created could prevent MXene from stacking. Furthermore, the inclusion of CoS_2 nanoparticles in the MXene material grades in a noteworthy rise in the total figure of active sites for the adsorption of metal ions on the electrode surface. This leads to a greater specific capacitance of 1320 F/g at current density of 1 A/g in 2M KOH electrolyte, and it has also achieved 78.4% of retention rate over 3000 cycles at a current density of 10 A/g. An example of ternary case, $NiCo_2S_4$/MXene was synthesized by Fu et al. [54] by electrostatically combining sisal-like $NiCo_2S_4$ (NCS) with delaminated MXene nanosheets. For this study, NCS was prepared thoroughly by mixing

Emerging Materials for Next Frontier Energy and Environment Applications   Materials Research Forum LLC
Materials Research Foundations 170 (2024) 145-167   https://doi.org/10.21741/9781644903292-7

$Na_2S.9H_2O$ with NiCo precursor. Then, the mixture was immediately transferred to a stainless-steel autoclave and subjected to a hydrothermal treatment at $160°C$ for 12 hrs. In order to obtain NCS, the solution was rinsed with DI water to remove unwanted moieties and kept for drying under vacuum at $60°C$. To eliminate any leftover cetyltrimethylammonium bromide (CTAB) the prepared NCS was combined with a solution of CTAB while stirring. Then the mixture was rinsed with DI water and, before being washed and dried at $70°C$, the CTAB treated NCS powder was mixed with an MXene solution in DI water. The zeta potential test revealed that the initially neutral NCS had a little positive charge of 4.47 mV, which increased to 28.44 mV after CTAB alteration. As a result, the equally distributed 2D MXene nanosheets on the surface of the NCS enable the composites to attain a 3D structure through spontaneous electrostatic contact between components with opposing charges. This material has strong conductivity and many edge sites for rapid reaction kinetics, is formed when 2D MXene nanosheets spread across a sisal-like structure. To measure the electrochemical performance, a three-electrode cell was set up with 3M KOH as the electrolyte and NCS/MXene as the working electrode. In a three-electrode cell arrangement with a 3M KOH electrolyte, NCS/MXene exhibits a 1028 C/g specific capacity at 1 A/g current density and a retention of 94.27% even after 5000 charge/discharge cycles. Furthermore, the fabricated NCS/MXene//AC pouch type asymmetric supercapacitor device in KOH electrolyte has symmetrical GCD curves, which implies that the device is highly reversible and displays high coulombic efficiency. The specific capacitance of the invented device is 171 F/g at 1 A/g and retails 89.5% of its capacitance after 5000 cycles, respectively.

4.3   Transition metal selenides/ MXene (TMSe's/MXene)

In comparison to TMSs, TMSe's showed significant improvements in electrical conductivity and dynamic edge facets. These materials on the other hand, possess greater thermal stability and good electrochemical activity, suggesting they might replace noble metal materials. Nevertheless, the processes of agglomeration, restacking and self-oxidation have posed serious issues for electrochemical applications. Consequently, functionalized MXene was integrated with diverse materials to facilitate the formation of nucleation or anchoring sites for active materials. This integration serves the purpose of preventing self-oxidation and restacking. Yet, there is a scarcity of research on MXene hybrid systems [55,56,57,58,59]. As far as we know, there has been no reports on the combination of metal selenides with these substances. As an example, $NiSe_2$ nanocrystals with MXene were reported by Jiang et.al [60] using a one-pot hydrothermal technique using $NiCl_2$ and selenium powder as a precursor. Herein, EDTA-2Na acts as a chelating agent to form a blue-colored solution followed by the addition of 1mL $Ti_3C_2T_x$ sheets. All these mixtures are transferred into a Teflon lined stainless steel autoclave and kept for the heat treatment at $180°C$ for 24 hrs. Finally, the obtained mixture was washed and dried at $60°C$ for 12 hrs. To Further examine the electrochemical performance, three electrodes set up was used and the analysis was done under 2M KOH electrolyte. These hybrid materials exhibit outstanding electrochemical performance having a high specific capacitance of about 531.2 F/g at a current density of 1 A/g. Further, the prolonged stability test demonstrates exceptional durability after 2000 cycles. Likewise, the hydrothermal approach of $MoSe_2$/MXene nanocomposite electrode was investigated by Arulkumar et al. [61] which displays ultra-thin nanosheet like morphology. These hybrid material displays the high specific capacitance of 1531.2 F/g at a current density of 1 A/g with 96.3% capacitance retention after 10000 cycles with 98% of coulombic efficiency. In contrast, an asymmetric device was fabricated by Samal et al. [62] by gathering $NiCoSe_2/Ti_3C_2T_x$ as the positive electrode whereas $Ti_3C_2T_x$ as the negative electrode with the utilization of a two-electrode setup under KOH/PVA gel electrolyte. These devices show a splendid energy density of 32.05 Wh/kg, whereas the power density is 0.2 kW/kg, which maintains a 15.39 Wh/kg energy density with a 0.92 kW/kg power density as well.

4.4   Transition metal tellurides/MXene (TMT's/MXene)

As per the literature, TMTs have a good gravimetric capacitance yet lack a lot of power rate capability as well as volumetric capacity. In this context, TMTs with a high conductive nature help the SCs device work at high speeds. These materials also possess the most metallic properties of all the other chalcogens owing to their lower electronegativity and bigger atoms than S and Se they can also hold more electrolyte ions which favours a faster diffusion rate. The paradigm shift in the field of electrochemical energy storage devices is possible as a consequence of our improved understanding of TMTs, which opens up a novel avenue for SCs [63]. Specifically, this includes the hybridization of binary or ternary TMTs with the 2D material known as MXene. According to our survey, only a few reports are available for this particular class of hybridized TMTs for SCs, some of which are VTe_2/ $Ti_3C_2T_x$ demonstrated by Raj et al. [64] which was synthesized by the simple hydrothermal method and displays 250 F/g of specific capacitance along with exceptional durability after a long cycle. Also, the asymmetric supercapacitor device was built with VTe_2/MXene and MoS_2/MXene as the positive and negative electrodes. Although, these device offers 46.3 Wh/kg energy density and displays the highest power density of 6400 W/kg, respectively. Thus, there might be more attention towards TMTs/MXene-based electrode materials for the high-performance supercapacitor applications. At last, we have discussed about the comparison of TMCs and TMCs/MXene with the previous literatures, as shown in Table 1.

Table 1 Comparison table for the electrochemical behaviour of featured TMCs and TMCs/MXene electrode materials for supercapacitors (Three and Two electrodes)

Electrode	Synthesis method	Electrolyte	Specific capacitance	Cyclic stability	Ref.
MoS_2	Exfoliation	1M H_2SO_4	700 F/cm^3	90% after 5000 cycles	[65]
WS_2	Hydrothermal	5M LiCl	93 F/cm^2	95.3% after 10000 cycles	[66]
Ni/MoS_2	Hydrothermal	0.5M H_2SO_4	305.9 F/g	87.4% after 5000 cycles	[67]
VS_2	Hydrothermal	6M KOH	155 A/g	99% after 5000 cycles	[68]
TiS_2/MoS_2	Hydrothermal	2M KOH	709 F/g	91% after 4000 cycles	[69]
MoS_2/TiS_2	Exfoliation	1M H_2SO_4	488.16 mF/cm^2	91% after 100000 cycles	[70]
ZrS_2	CVT	1M H_2SO_4	18.8 F/g	88% after 5000 cycles	[71]
NbS_2	Colloidal	1M KCl	221.4 F/g	78.9% after 10000 cycles	[72]
TaS_2	Hydrothermal	6M KOH	502 F/cm^3	91% after 5000 cycles	[73]
NiS_2/MoS_2/G	Solvothermal	2M KOH	2379 F/g	60.7% at 100 A/g	[74]

Electrode	Synthesis Method	Current Collector	Electrolyte	Specific Capacitance	Cyclic Stability	Ref.
$Cu_{0.5}Co_{0.5}Se_2$	Hydrothermal	Ni Foam	2 M KOH	1695 F/g @ 1 A/g	94.9% (10000)	[75]
$Ti_3C_2T_x/$ $MoSn_2Se_4$	Hydrothermal	Carbon Cloth	3 M KOH	212 F/g @1 A/g	88% (10000)	[76]
NiMnSe	Hydrothermal	Ni Foam	6 M KOH	1802.9 F/g @0.5 A/g	85% (3000)	[77]
$Ti_3C_2T_x/NiSe_2$	Hydrothermal	Ni Foam	2 M KOH	531.2 F/g @ 1 A/g	66% (1000)	[60]
$CoSe_2$	Microwave	Ni Foam	6 M KOH	333 F/g @ 1 A/g	100.97% (25000)	[22]
$Ti_3C_2T_x/WSe_2$	Hydrothermal	Ni Foam	6 M KOH	840 F/g @ 2 A/g	92% (5000)	[78]
Co-Fe @ Te	Wet Chemical	Ni Foam	4 M KOH	1119.2 F/g @ 6 A/g	92.2% (2000)	[79]
CoTe	Hydrothermal	Ni Foam	3.5 M KOH	170 C/g @ 0.5 A/g	99% (5000)	[32]
NiCoTe	Hydrothermal	Graphite Paper	3.5 M KOH	433 C/g @0.5 A/g	100% (5000)	[80]
$NiTe_2/ CoTe_2$	Hydrothermal	Ni Foam	6 M KOH	1388.9 C/g @1A/g	83.45% (6500)	[35]
$Ti_3C_2T_x/VTe_2$	Hydrothermal	Ni Foam	0.5 M K_2SO_4	250 F/g @ 0.25 A/g	83.5% (7000)	[64]

Electrode	Electrolyte	Specific Capacitance	Cyclic Stability	Energy Density	Power Density	Ref.
$Ti_3C_2T_x//$ $Cu_{0.5}Co_{0.5}Se_2$	PVA- KOH	236.72 F/g	91.1% (10000)	84.17 Wh/kg	14.95 kW/kg	[75]
$Ti_3C_2T_x/$ $MoSn_2Se_4//$ rGO	1M KOH	120.2 F/g	97% (10000)	42.7 Wh/kg	1598.6 W/kg	[76]
$Ti_3C_2T_x/NiCoSe_2$	PVA- KOH	160.25 F/g	95.67% (10000)	32.05 Wh/kg	0.2 kW/kg	[62]
$CoSe_2//AC$	6M KOH	56.7 F/g	93.3% (20000)	18.9 Wh/kg	387 W/kg	[81]
$Ti_3C_2T_x/WSe_2$ $// Ti_3C_2T_x/WSe_2$	6M KOH	246 F/g	97% (5000)	6 Wh/kg	0.6 k W/kg	[78]
Co-Fe @ Te//AC	4M KOH	179.2 F/g	73% (4000)	62.1 Wh/kg	1138.2 W/kg	[79]
Co Te//AC	3.5M KOH	183 C/g	85% (10000)	40.7 Wh/kg	800 W/kg	[32]
NiCoTe//AC	3.5M KOH	382 C/g	91% (25000)	43 Wh/kg	905 W/kg	[80]
$Ti_3C_2T_x/VTe_2//$ $Ti_3C_2T_x/MoS_2$	0.5M K_2SO_4	130 F/g	87% (7000)	46.3 Wh/kg	400 W/kg	[64]

Conclusion

In this chapter, we discussed the need for storage, renewable energy sources, and the advancement of materials for SCs including TMCs, MXene and their composite characteristics, as well as the methodical and thorough discussion of their consequences for high-performance SCs. After providing a short overview of energy conservation, the chapter explores the types and overview of

SCs, the development of TMCs materials, and deals with their previously reported articles. Following that, we discussed the novel class of 2D material called MXene and their composites with TMCs such as TMSs, TMSe's, and TMTs, respectively. Though this chapter aims to provide a brief summary of experimental research on 2D materials, especially TMCs, MXene, and its derivatives, for their capacitive features in high energy and high-power SCs. We highlight recent findings that enhance the use of MXene in the production of SCs by controlling their size, shape, and unique morphology. This current discussion also describes the charge storage mechanisms employed by pure TMCs, MXene, and their hybrids as well. Added to that, the latest advances in heterostructure technology as well as the synthesis procedure advancements that have been connected to further enhance the supercapacitor performance. Finally, we address the remaining technical concerns, discuss emerging technologies, and explore the potential applications of SCs in the energy and power domains TMCs/MXene hybrids have an array of energy storage applications, but more research is needed to determine optimal structural features and synthesis approaches. In the meantime, many TMCs/MXene hybrid materials have been put forward as electrode materials for energy related applications. Due to limited resource availability, theoretical considerations for such a class of materials sometimes oversimplify and idealize models, resulting in an unduly optimistic view. To get optimal performance, it is crucial to create and evaluate well-defined TMCs/MXene hybrids. To understand completely the surface chemistry and the electrochemical reaction processes of TMCs/MXene hybrids, various characterizations and analytical techniques are recommended. Finally, this chapter strives to promote research into diverse TMCs/MXene hybrid pairings. Thus, paving the way for next-generation energy alternatives.

References

[1] S.E.E.Küçükkambak, Environmental Solutions and Alternative Policies to Energy Crises on the Basis of Renewable Energy Production and the Global RenewableEnergyMarket,. https://doi.org/10.4018/979-8-3693-0440-2.CH003.

[2] H.S. Magar, A.M. Mansour, A.B.A. Hammad, Advancing energy storage and supercapacitor applications through the development of Li+-doped $MgTiO_3$ perovskite nano-ceramics, Sci Reports 2024 141 14 (2024) 1–18. https://doi.org/10.1038/s41598-024-52262-6.

[3] K.V.G. Raghavendra, R. Vinoth, K. Zeb, C.V.V. Muralee Gopi, S. Sambasivam, M.R. Kummara, I.M. Obaidat, H.J. Kim, An intuitive review of supercapacitors with recent progress and novel device applications, J Energy Storage 31 (2020) 101652. https://doi.org/10.1016/J.EST.2020.101652.

[4] Z.A. Sheikh, P.K. Katkar, H. Kim, S. Rehman, K. Khan, V.D. Chavan, R. Jose, M.F. Khan, D. kee Kim, Transition metal chalcogenides, MXene, and their hybrids: An emerging electrochemical capacitor electrodes, J Energy Storage 71 (2023) 107997. https://doi.org/10.1016/J.EST.2023.107997.

[5] T.E. Balaji, H. Tanaya Das, T. Maiyalagan, Recent Trends in Bimetallic Oxides and Their Composites as Electrode Materials for Supercapacitor Applications, ChemElectroChem8(2021)1723–1746. https://doi.org/10.1002/CELC.202100098.

[6]S.G. Krishnan, A. Arunachalam, P. Jagadish, M. Khalid, 2D Materials for Supercapacitor and Supercapattery Applications, ACS Symp Ser 1353 (2020) 33–47. https://doi.org/10.1021/BK-2020-1353.CH002.

[7]Z.S. Iro, C. Subramani, S.S. Dash, A Brief Review on Electrode Materials for Supercapacitor, Int J Electrochem Sci 11 (2016) 10628–10643. https://doi.org/10.20964/2016.12.50.

[8]J. Xie, P. Yang, Y. Wang, T. Qi, Y. Lei, C.M. Li, Puzzles and confusions in supercapacitor and battery: Theory and solutions, J Power Sources 401 (2018) 213–223. https://doi.org/10.1016/J.JPOWSOUR.2018.08.090.

[9]X. Chen, R. Paul, L. Dai, Carbon-based supercapacitors for efficient energy storage, Natl Sci Rev 4 (2017) 453–489. https://doi.org/10.1093/NSR/NWX009.

[10]   J. Cherusseri, N. Choudhary, K. Sambath Kumar, Y. Jung, J. Thomas, Recent trends in transition metal dichalcogenide based supercapacitor electrodes, Nanoscale Horizons 4 (2019) 840–858. https://doi.org/10.1039/C9NH00152B.

[11]   D. Monga, S. Sharma, N.P. Shetti, S. Basu, K.R. Reddy, T.M. Aminabhavi, Advances in transition metal dichalcogenide-based two-dimensional nanomaterials, Mater Today Chem 19 (2021) 100399. https://doi.org/10.1016/J.MTCHEM.2020.100399.

[12]   Y. Liu, Y. Li, H. Kang, T. Jin, L. Jiao, Design, synthesis, and energy-related applications of metal sulfides, Mater Horizons 3 (2016) 402–421. https://doi.org/10.1039/C6MH00075D.

[13]   P. Kulkarni, S.K. Nataraj, R.G. Balakrishna, D.H. Nagaraju, M. V. Reddy, Nanostructured binary and ternary metal sulfides: synthesis methods and their application in energy conversion and storage devices, J Mater Chem A 5 (2017) 22040–22094. https://doi.org/10.1039/C7TA07329A.

[14]   S. Chandrasekaran, L. Yao, L. Deng, C. Bowen, Y. Zhang, S. Chen, Z. Lin, F. Peng, P. Zhang, Recent advances in metal sulfides: From controlled fabrication to electrocatalytic, photocatalytic and photoelectrochemical water splitting and beyond, Chem Soc Rev 48 (2019) 4178–4280. https://doi.org/10.1039/C8CS00664D.

[15]   A. Das, B. Raj, M. Mohapatra, S.M. Andersen, S. Basu, Performance and future directions of transition metal sulfide-based electrode materials towards supercapacitor/supercapattery, Wiley Interdiscip Rev Energy Environ 11 (2022). https://doi.org/10.1002/WENE.414.

[16]   S. Wei, R. Zhou, G. Wang, Enhanced Electrochemical Performance of Self-Assembled Nanoflowers of MoS2 Nanosheets as Supercapacitor Electrode Materials,ACS Omega 4 (2019). https://doi.org/10.1021/ACSOMEGA.9B01058.

[17]   C. Nagaraju, C.V.V.M. Gopi, J.W. Ahn, H.J. Kim, Hydrothermal synthesis of MoS2 and WS2 nanoparticles for high-performance supercapacitor applications, New J Chem 42 (2018) 12357–12360. https://doi.org/10.1039/C8NJ02822B.

[18]   T.N.V. Krishna, P. Himasree, K.V.G. Raghavendra, S.S. Rao, N.B. Kundakarla, D. Punnoose, H.J. Kim, Hydrothermal synthesis of layered CoS@WS2 nanocomposite as a potential electrode for high-performance supercapacitor applications, J Mater Sci Mater Electron 31 (2020) 16290–16298. https://doi.org/10.1007/S10854-020-04177-X.

[19]   S. Wang, P. Zhang, C. Liu, Synthesis of hierarchical bimetallic sulfide NiCo2S4 for high-performance supercapacitors, Colloids Surfaces A Physicochem Eng Asp 616 (2021) 126334. https://doi.org/10.1016/J.COLSURFA.2021.126334.

[20]   X. Liu, D. Zhang, Y. Ma, G. Li, X. Yuan, Y. Huang, G. Liu, M. Guo, W. Zheng, M2+-Doped Nickel Selenide Nanoflowers (M = Co, Cu, and Zn) for Supercapacitors, ACS Appl Nano Mater (2024). https://doi.org/10.1021/ACSANM.4C00313/SUPPL_FILE/AN4C00313_SI_001.PDF.

[21]   M. Sajjad, M. Amin, M.S. Javed, M. Imran, W. Hu, Z. Mao, W. Lu, Recent trends in transition metal diselenides (XSe2: X = Ni, Mn, Co) and their composites for high energy faradic supercapacitors, J Energy Storage 43 (2021) 103176. https://doi.org/10.1016/J.EST.2021.103176.

[22]   S. Liu, S. Sarwar, J. Wang, H. Zhang, T. Li, J. Luo, X. Zhang, The microwave synthesis of porous CoSe2 nanosheets for super cycling performance supercapacitors, J Mater Chem C 9 (2021) 228–237. https://doi.org/10.1039/D0TC04718J.

[23]   J. Yang, Z. Sun, J. Wang, J. Zhang, Y. Qin, J. You, L. Xu, Hierarchical NiSe2 spheres composed of tiny nanoparticles for high performance asymmetric supercapacitors, CrystEngComm 21 (2019) 994–1000. https://doi.org/10.1039/C8CE01805G.

[24]   Y. Liu, W. Li, X. Chang, H. Chen, X. Zheng, J. Bai, Z. Ren, MoSe2 nanoflakes-decorated vertically aligned carbon nanotube film on nickel foam as a binder-free supercapacitor electrode with high rate capability, J Colloid Interface Sci 562 (2020) 483–492. https://doi.org/10.1016/J.JCIS.2019.11.089.

[25]   P.A. Shinde, N.R. Chodankar, M.A. Abdelkareem, S.J. Patil, Y.K. Han, K. Elsaid, A.G. Olabi, All Transition Metal Selenide Composed High-Energy Solid-State Hybrid Supercapacitor, Small 18 (2022). https://doi.org/10.1002/SMLL.202200248.

[26]   C.V.V. Muralee Gopi, A.E. Reddy, H.J. Kim, Wearable superhigh energy density supercapacitors using a hierarchical ternary metal selenide composite of CoNiSe2 microspheres decorated with CoFe2Se4 nanorods, J Mater Chem A 6 (2018) 7439–7448. https://doi.org/10.1039/C8TA01141A.

[27]   K. Wang, X. Wang, Z. Li, B. Yang, M. Ling, X. Gao, J. Lu, Q. Shi, L. Lei, G. Wu, Y. Hou, Designing 3d dual transition metal electrocatalysts for oxygen evolution reaction in alkaline electrolyte: Beyond oxides, Nano Energy 77 (2020) 105162. https://doi.org/10.1016/J.NANOEN.2020.105162.

[28]   T. Zhang, J. Li, R. Bi, J. Song, L. Du, T. Li, H. Zhang, Q. Guo, J. Luo, One-step microwave synthesis of in situ grown NiTe nanosheets for solid-state asymmetric supercapacitors and oxygen evolution reaction, J Alloys Compd 909 (2022) 164786. https://doi.org/10.1016/J.JALLCOM.2022.164786.

[29]   B. Ye, M. Huang, L. Fan, J. Lin, J. Wu, Co ions doped NiTe electrode material for asymmetric supercapacitor application, J Alloys Compd 776 (2019) 993–1001. https://doi.org/10.1016/J.JALLCOM.2018.10.358.

[30]   M. Manikandan, K. Subramani, M. Sathish, S. Dhanuskodi, NiTe Nanorods as Electrode Material for High Performance Supercapacitor Applications, ChemistrySelect 3 (2018) 9034–9040. https://doi.org/10.1002/SLCT.201801421.

[31]   B. Ye, M. Huang, Q. Bao, S. Jiang, J. Ge, H. Zhao, L. Fan, J. Lin, J. Wu, Construction of NiTe/NiSe Composites on Ni Foam for High-Performance Asymmetric Supercapacitor, ChemElectroChem 5 (2018) 507–514. https://doi.org/10.1002/CELC.201701033.

[32]   M. Manikandan, K. Subramani, M. Sathish, S. Dhanuskodi, Hydrothermal synthesis of cobalt telluride nanorods for a high performance hybrid asymmetric supercapacitor, RSC Adv 10 (2020) 13632–13641. https://doi.org/10.1039/C9RA08692G.

[33]   M. Manikandan, K. Subramani, S. Dhanuskodi, M. Sathish, One-Pot Hydrothermal Synthesis of Nickel Cobalt Telluride Nanorods for Hybrid Energy Storage Systems, Energy and Fuels 35 (2021) 12527–12537. https://doi.org/10.1021/ACS.ENERGYFUELS.1C00351.

[34]   P. Bhol, P.B. Jagdale, A.H. Jadhav, M. Saxena, A.K. Samal, All-Solid-State Supercapacitors Based on Cobalt Magnesium Telluride Microtubes Decorated with Tellurium Nanotubes, ChemSusChem 17 (2024) e202301009. https://doi.org/10.1002/CSSC.202301009.

[35]   M. Molaei, G.R. Rostami, A.M. Zardkhoshoui, S.S.H. Davarani, In situ tellurization strategy for crafting nickel ditelluride/cobalt ditelluride hierarchical nanostructures: A leap forward in hybrid supercapacitor electrode materials, J Colloid Interface Sci 653 (2024) 1683–1693. https://doi.org/10.1016/J.JCIS.2023.10.012.

[36]   L. Zhu, P. Xu, B. Chang, J. Ning, T. Yan, Z. Yang, H. Lu, Hierarchical Structure by Self-sedimentation of Liquid Metal for Flexible Sensor Integrating Pressure Detection and Triboelectric Nanogenerator, Adv Funct Mater (2024) 2400363. https://doi.org/10.1002/ADFM.202400363.

[37]   T. Yuan, Y. Li, S. Chen, X. Ren, P. Zhao, X. Zhao, H. Shan, Microstructural evolution and mechanisms affecting the mechanical properties of wire arc additively manufactured Al-Zn-Mg-Cu alloy reinforced with high-entropy alloy particles, J Alloys Compd 992 (2024) 174582. https://doi.org/10.1016/J.JALLCOM.2024.174582.

[38]   S.E. Moosavifard, A. Mohammadi, M. Ebrahimnejad Darzi, A. Kariman, M.M. Abdi, G. Karimi, A facile strategy to synthesis graphene-wrapped nanoporous copper-cobalt-selenide hollow spheres as an efficient electrode for hybrid supercapacitors, Chem Eng J 415 (2021) 128662. https://doi.org/10.1016/J.CEJ.2021.128662.

[39]   H. Jiang, Z. Wang, Q. Yang, M. Hanif, Z. Wang, L. Dong, M. Dong, A novel MnO2/Ti3C2Tx MXene nanocomposite as high performance electrode materials for flexible supercapacitors, Electrochim Acta 290 (2018) 695–703. https://doi.org/10.1016/J.ELECTACTA.2018.08.096.

[40]   C. Eames, M.S. Islam, Ion intercalation into two-dimensional transition-metal carbides: Global screening for new high-capacity battery materials, J Am Chem Soc 136 (2014) 16270–16276. https://doi.org/10.1021/JA508154E/SUPPL_FILE/JA508154E_SI_001.PDF.

[41]   M. Ghidiu, M.R. Lukatskaya, M.Q. Zhao, Y. Gogotsi, M.W. Barsoum, Conductive two-dimensional titanium carbide "clay" with high volumetric capacitance., Nature 516 (2014) 78–81. https://doi.org/10.1038/NATURE13970.

[42]   M.R. Lukatskaya, O. Mashtalir, C.E. Ren, Y. Dall'Agnese, P. Rozier, P.L. Taberna, M. Naguib, P. Simon, M.W. Barsoum, Y. Gogotsi, Cation intercalation and high volumetric capacitance of two-dimensional titanium carbide, Science 341 (2013) 1502–1505. https://doi.org/10.1126/SCIENCE.1241488.

[43]   X. Wang, C. Garnero, G. Rochard, D. Magne, S. Morisset, S. Hurand, P. Chartier, J. Rousseau, T. Cabioc'H, C. Coutanceau, V. Mauchamp, S. Célérier, A new etching environment (FeF3/HCl) for the synthesis of two-dimensional titanium carbide MXenes: a route towards selective reactivity vs. water, J Mater Chem A 5 (2017) 22012–22023. https://doi.org/10.1039/C7TA01082F.

[44]   P. Urbankowski, B. Anasori, T. Makaryan, D. Er, S. Kota, P.L. Walsh, M. Zhao, V.B. Shenoy, M.W. Barsoum, Y. Gogotsi, Synthesis of two-dimensional titanium nitride Ti4N3 (MXene), Nanoscale 8 (2016) 11385–11391. https://doi.org/10.1039/C6NR02253G.

[45]   Z. Ali, T. Zhang, M. Asif, L. Zhao, Y. Yu, Y. Hou, Transition metal chalcogenide anodes for sodium storage, Mater Today 35 (2020) 131–167. https://doi.org/10.1016/J.MATTOD.2019.11.008.

[46]   H. Guo, J. Zhang, F. Yang, M. Wang, T. Zhang, Y. Hao, W. Yang, Sandwich-like porous MXene/Ni3S4/CuS derived from MOFs as superior supercapacitor electrode, J Alloys Compd 906 (2022) 163863. https://doi.org/10.1016/J.JALLCOM.2022.163863.

[47]   Y. Luo, Y. Tian, Y. Tang, X. Yin, W. Que, 2D hierarchical nickel cobalt sulfides coupled with ultrathin titanium carbide (MXene) nanosheets for hybrid supercapacitors, J Power Sources 482 (2021) 228961. https://doi.org/10.1016/J.JPOWSOUR.2020.228961.

[48]   M.S. Javed, X. Zhang, S. Ali, A. Mateen, M. Idrees, M. Sajjad, S. Batool, A. Ahmad, M. Imran, T. Najam, W. Han, Heterostructured bimetallic–sulfide@layered Ti3C2Tx–MXene as a synergistic electrode to realize high-energy-density aqueous hybrid-supercapacitor, Nano Energy 101 (2022) 107624. https://doi.org/10.1016/J.NANOEN.2022.107624.

[49]   J.Q. Qi, M.Y. Huang, C.Y. Ruan, D.D. Zhu, L. Zhu, F.X. Wei, Y.W. Sui, Q.K. Meng, Construction of CoNi2S4 nanocubes interlinked by few-layer Ti3C2Tx MXene with high performance for asymmetric supercapacitors, Rare Met 41 (2022) 4116–4126. https://doi.org/10.1007/S12598-022-02167-Y/FIGURES/8.

[50]   J.Q. Qi, C.C. Zhang, H. Liu, L. Zhu, Y.W. Sui, X.J. Feng, W.Q. Wei, H. Zhang, P. Cao, MXene-wrapped ZnCo2S4 core–shell nanospheres via electrostatic self-assembly as positive electrode materials for asymmetric supercapacitors, Rare Met 41 (2022) 2633–2644. https://doi.org/10.1007/S12598-021-01956-1/TABLES/1.

[51]   Y. Shen, H. Jiang, Z. Lu, G. Li, Z. Wang, J. Zhang, Facile decoration of two-dimensional Ti3C2T x nanoplates with CuS nanoparticles via a facile in situ synthesis strategy at room temperature for superhigh specific capacitance of supercapacitors, Nanotechnology 33 (2021). https://doi.org/10.1088/1361-6528/AC30F2.

[52]   W. Zhang, J. Qi, T. Cao, Z. Lei, Y. Ma, H. Liu, L. Zhu, X. Feng, W. Wei, H. Zhang, A facile method synthesizing marshmallow ZnS grown on Ti3C2 MXene for high-performance asymmetric supercapacitors, J Energy Storage 50 (2022) 104652. https://doi.org/10.1016/J.EST.2022.104652.

[53]   H. Liu, R. Hu, J. Qi, Y. Sui, Y. He, Q. Meng, F. Wei, Y. Ren, Y. Zhao, W. Wei, One-Step Synthesis of Nanostructured CoS2 Grown on Titanium Carbide MXene for High-Performance Asymmetrical Supercapacitors, Adv Mater Interfaces 7 (2020) 1901659. https://doi.org/10.1002/ADMI.201901659.

[54]   J. Fu, L. Li, D. Lee, J.M. Yun, B.K. Ryu, K.H. Kim, Enhanced electrochemical performance of Ti3C2Tx MXene film based supercapacitors in H2SO4/KI redox additive electrolyte, Appl Surf Sci 504 (2020) 144250. https://doi.org/10.1016/J.APSUSC.2019.144250.

[55]   H. Huang, J. Cui, G. Liu, R. Bi, L. Zhang, Carbon-Coated MoSe 2 /MXene Hybrid Nanosheets for Superior Potassium Storage, ACS Nano 13 (2019) 3448–3456. https://doi.org/10.1021/ACSNANO.8B09548/SUPPL_FILE/NN8B09548_SI_001.PDF.

[56]   T. Kshetri, D.T. Tran, H.T. Le, D.C. Nguyen, H. Van Hoa, N.H. Kim, J.H. Lee, Recent advances in MXene-based nanocomposites for electrochemical energy storage applications, Prog Mater Sci 117 (2021) 100733. https://doi.org/10.1016/J.PMATSCI.2020.100733.

[57]   T. Chen, S. Li, J. Wen, P. Gui, G. Fang, Metal-Organic Framework Template Derived Porous CoSe2 Nanosheet Arrays for Energy Conversion and Storage, ACS Appl Mater Interfaces 9 (2017) 35927–35935. https://doi.org/10.1021/ACSAMI.7B12403/SUPPL_FILE/AM7B12403_SI_002.PDF.

[58]   Z. Wang, Q. Li, Y. Chen, B. Cui, Y. Li, F. Besenbacher, M. Dong, The ambipolar transport behavior of WSe2 transistors and its analogue circuits, NPG Asia Mater 2018 108 10 (2018) 703–712. https://doi.org/10.1038/s41427-018-0062-1.

[59]   Z. Wang, Q. Li, F. Besenbacher, M. Dong, Facile Synthesis of Single Crystal PtSe2 Nanosheets for Nanoscale Electronics, Adv Mater 28 (2016) 10224–10229. https://doi.org/10.1002/ADMA.201602889.

[60]   H. Jiang, Z. Wang, Q. Yang, L. Tan, L. Dong, M. Dong, Ultrathin Ti3C2Tx (MXene) Nanosheet-Wrapped NiSe2 Octahedral Crystal for Enhanced Supercapacitor Performance and Synergetic Electrocatalytic Water Splitting, Nano-Micro Lett 11 (2019) 1–14. https://doi.org/10.1007/S40820-019-0261-5/FIGURES/4.

[61]   C. Arulkumar, R. Gandhi, S. Vadivel, Ultra-thin nanosheets of Ti3C2Tx MXene/MoSe2 nanocomposite electrode for asymmetric supercapacitor and electrocatalytic water splitting, Electrochim Acta 462 (2023) 142742. https://doi.org/10.1016/J.ELECTACTA.2023.142742.

[62]   R. Samal, P. Mane, S. Ratha, B. Chakraborty, C.S. Rout, Rational Design of Dynamic Bimetallic NiCoSe2/2D Ti3C2T xMXene Hybrids for a High-Performance Flexible Supercapacitor and Hydrogen Evolution Reaction, Energy and Fuels 36 (2022) 15066–15079. https://doi.org/10.1021/ACS.ENERGYFUELS.2C02808/SUPPL_FILE/EF2C02808_SI_001. PDF.

[63]   A.J. Khan, M. Sajjad, S. Khan, M. Khan, A. Mateen, S.S. Shah, N. Arshid, L. He, Z. Ma, L. Gao, G. Zhao, Telluride-Based Materials: A Promising Route for High Performance Supercapacitors, Chem Rec 24 (2024) e202300302. https://doi.org/10.1002/TCR.202300302.

[64]   K.A. Sree Raj, N. Barman, S. Radhakrishnan, R. Thapa, C.S. Rout, Hierarchical architecture of the metallic VTe2/Ti3C2Tx MXene heterostructure for supercapacitor applications, J Mater Chem A 10 (2022) 23590–23602. https://doi.org/10.1039/D2TA05904E.

[65]   M. Acerce, D. Voiry, M. Chhowalla, Metallic 1T phase MoS2 nanosheets as supercapacitor electrode materials, Nat Nanotechnol 2015 104 10 (2015) 313–318. https://doi.org/10.1038/nnano.2015.40.

[66]   S. Liu, Y. Zeng, M. Zhang, S. Xie, Y. Tong, F. Cheng, X. Lu, Binder-free WS2 nanosheets with enhanced crystallinity as a stable negative electrode for flexible asymmetric supercapacitors, J Mater Chem A 5 (2017) 21460–21466. https://doi.org/10.1039/C7TA07009H.

[67]   S.B. Saseendran, A. Ashok, A.S. Asha, Flexible and binder-free supercapacitor electrode with high mass loading using transition metal doped MoS2 nanostructures, J Alloys Compd 968 (2023) 172131. https://doi.org/10.1016/J.JALLCOM.2023.172131.

[68]   T.M. Masikhwa, F. Barzegar, J.K. Dangbegnon, A. Bello, M.J. Madito, D. Momodu, N. Manyala, Asymmetric supercapacitor based on VS2 nanosheets and activated carbon materials, RSC Adv 6 (2016) 38990–39000. https://doi.org/10.1039/C5RA27155J.

[69]   G. Nabi, W. Ali, M. Tanveer, T. Iqbal, M. Rizwan, S. Hussain, Robust synergistic effect of TiS2/MoS2 hierarchal micro-flowers composite realizing enhanced electrochemical performance, J Energy Storage 58 (2023) 106316. https://doi.org/10.1016/J.EST.2022.106316.

[70]   A. Panagiotopoulos, G. Nagaraju, S. Tagliaferri, C. Grotta, P.C. Sherrell, M. Sokolikova, G. Cheng, F. Iacoviello, K. Sharda, C. Mattevi, 3D printed inks of two-dimensional semimetallic MoS2/TiS2 nanosheets for conductive-additive-free symmetric supercapacitors, J Mater Chem A 11 (2023) 16190–16200. https://doi.org/10.1039/D3TA02508J.

[71]   M. Habib, S. Ullah, F. Khan, M.I. Rafiq, A. Salem Balobaid, T. Alshahrani, Z. Muhammad, Supercapacitor electrodes based on single crystal layered ZrX2 (X = S, Se) using chemical vapor transport method, Mater Sci Eng B 298 (2023) 116904. https://doi.org/10.1016/J.MSEB.2023.116904.

[72]   W. Li, X. Wei, H. Dong, Y. Ou, S. Xiao, Y. Yang, P. Xiao, Y. Zhang, Colloidal Synthesis of NbS2 Nanosheets: From Large-Area Ultrathin Nanosheets to Hierarchical Structures, Front Chem 8 (2020) 533530. https://doi.org/10.3389/FCHEM.2020.00189/BIBTEX.

[73]   M. Zhang, Y. He, D. Yan, H. Xu, A. Wang, Z. Chen, S. Wang, H. Luo, K. Yan, Multifunctional 2H-TaS2 nanoflakes for efficient supercapacitors and electrocatalytic evolution of hydrogen and oxygen, Nanoscale 11 (2019) 22255–22260. https://doi.org/10.1039/C9NR07564J.

[74]   X. Yang, J. Mao, H. Niu, Q. Wang, K. Zhu, K. Ye, G. Wang, D. Cao, J. Yan, NiS2/MoS2 mixed phases with abundant active edge sites induced by sulfidation and graphene introduction towards high-rate supercapacitors, Chem Eng J 406 (2021) 126713. https://doi.org/10.1016/J.CEJ.2020.126713.

[75]   Y. Abu Dakka, J. Balamurugan, R. Balaji, N.H. Kim, J.H. Lee, Advanced Cu0.5Co0.5Se2 nanosheets and MXene electrodes for high-performance asymmetric supercapacitors, Chem Eng J 385 (2020) 123455. https://doi.org/10.1016/J.CEJ.2019.123455.

[76]   S. De, S. Roy, G.C. Nayak, MoSn2Se4-decorated MXene/functionalized RGO nanohybrid for ultrastable supercapacitor and oxygen evolution catalyst, Mater Today Nano 22 (2023) 100337. https://doi.org/10.1016/J.MTNANO.2023.100337.

[77]   A.M. Abuelftooh, M.G. Fayed, S.Y. Attia, Y.F. Barakat, N.S. Tantawy, S.G. Mohamed, A three-dimensional directly grown hierarchical graces-like Nickel Manganese Selenide for high-performance Li-ion battery and supercapacitor electrodes, Mater Today Chem 26 (2022) 101187. https://doi.org/10.1016/J.MTCHEM.2022.101187.

[78]   S. Hussain, D. Vikraman, M.T. Mehran, M. Hussain, G. Nazir, S.A. Patil, H.S. Kim, J. Jung, Ultrasonically derived WSe2 nanostructure embedded MXene hybrid composites for supercapacitors and hydrogen evolution reactions, Renew Energy 185 (2022) 585–597. https://doi.org/10.1016/J.RENENE.2021.12.065.

[79]   P. Bhol, S. Swain, A. Altaee, M. Saxena, A.K. Samal, Cobalt–iron decorated tellurium nanotubes for high energy density supercapacitor, Mater Today Chem 24 (2022) 100871. https://doi.org/10.1016/J.MTCHEM.2022.100871.

[80]   M. Manikandan, K. Subramani, S. Dhanuskodi, M. Sathish, One-Pot Hydrothermal Synthesis of Nickel Cobalt Telluride Nanorods for Hybrid Energy Storage Systems, Energy and Fuels 35 (2021) 12527–12537. https://doi.org/10.1021/ACS.ENERGYFUELS.1C00351/SUPPL_FILE/EF1C00351_SI_001.PDF.

[81]   S. Liu, S. Sarwar, J. Wang, H. Zhang, T. Li, J. Luo, X. Zhang, The microwave synthesis of porous CoSe2 nanosheets for super cycling performance supercapacitors, J Mater Chem C 9 (2021) 228–237. https://doi.org/10.1039/D0TC04718J.

Keyword Index

2D Materials	21, 145
Activation	60
Biomass	60
Carbon Derived from MOF	94
Carbon Electrodes	60
Carbonization	60
CO2 Reduction	21
Electrocatalysis	41
Electrode Material	94
Energy Storage	1
Energy Storage	117
GO/rGO Nanocomposites	1
Green Energy	41
Hydrogen Evolution Reaction	41
Metal Chalcogenides based Supercapacitors	1
Metal-Organic Framework	94
MXene	117, 145
Nanocomposites	145
Pollutant Degradation	21
Pristine MOF	94
Semiconductor Photocatalyst	21
Supercapacitor	60, 94, 117, 145
Ternary Metal Chalcogenides	145
Transition Metal Oxides	117
Tunable Bandgap	21
Two-Dimensional Materials	117
Water Splitting Reactions	41
Water Splitting	21

About the Editors

Dr. Alagarsamy Pandikumar is currently working as Senior Scientist in CSIR-Central Electrochemical Research Institute, Karaikudi, India. He obtained his Ph.D. in Chemistry (2014) from the Madurai Kamaraj University, Madurai. He successfully completed his post-doctoral fellowship tenure (2014-2016) at the University of Malaya, Malaysia under a High Impact Research Grant. He is in the top 2% of scientists list in the world database released by Stanford University, US 2024 (based on Scopus citation records). Moreover He is recognized as the World's Top 2% Scientists in the Single-Year Category and also in the Career Achievement Category. He has continuously achieved this record for the last four years (2021, 2022, 2023 & 2024). Further he has been awarded "RSC Affiliate member" by Royal Society of Chemistry, UK (2021). Recently He has received the prestigious Young Scientist Award for 2023 (Saraswathy Srinivasan Prize) from 'The Academy of Sciences, Chennai, Tamil Nadu' for his pioneering contribution to solar energy conversion. In addition he is holding an Assistant Professor position at Faculty of Chemical Sciences, Academy of Scientific and Innovative Research (AcSIR), Ghaziabad. Hs is a Life Member - Society for Advancement of Electrochemical Science & Technology (SAEST). His current research involves the development of novel functional materials for photocatalysis, photoelectrocatalysis, dye-sensitized solar cells, perovskite solar cells and water splitting applications. His results were documented in 160 peer-reviewed journals including 14 review articles; and also have 8600 citations with the h−index of 55.

Dr. Mani Alagiri is currently working as an Associate Professor at SRM Institute of Science and Technology, Chengalpattu, India. He completed his Ph.D. in the field of physics from SRM University, Chengalpattu in 2012 and subsequently, he did his postdoctoral fellowship at the University of Malaya, Malaysia in the year 2013 to 2014. He has been listed among the top 2% of scientists globally in the world database released by the Stanford University in both 2023 and 2024, based on Scopus Citation Records. His research focuses on the development of novel materials for environmental sustainability, particularly in the fields of environmental applications, energy conversion and energy storage such as degradation of textile and pharmaceutical wastes, water-splitting, and supercapacitors. Dr. Mani Alagiri is a life member of the Indian Association of Physics and the Society for Materials Chemistry. He has published about 47 peer-reviewed articles with over 1,546 citations and holds an h-index of 21, showcasing his influence in the scientific community.

www.ingramcontent.com/pod-product-compliance
Lightning Source LLC
Chambersburg PA
CBHW071235210326
41597CB00016B/2061